BETWEEN THE LINES

Concepts in Sound System Design and Alignment

Michael Lawrence

Foreword by Bob McCarthy

PRECISION AUDIO PRESS

Copyright © 2022 by Michael Lawrence, All Rights Reserved

No part of this book may be reprinted or reproduced or utilized in any form or by any electronic, mechanical, or other means, now known or hereafter invented, including photocopying and recording, or in any information storage or retrieval system, without permission in writing from the publisher.

Although every precaution has been taken to verify the accuracy of the information contained herein, the author and publisher assume no responsibility for any errors or omissions. No liability is assumed for damages that may result from the use of information contained within.

Trademark notice: Product or corporate names may be trademarks or registered trademarks, and are used only for identification and explanation without intent to infringe.

Precision Audio Press
New York, USA
www.precisionaudioservices.com

Print ISBN-13: 979-8-2180075-3-9
EBook ISBN-13: 979-8-218-04896-9
LCCN: 2020918289
BISAC: TEC001000 Technology & Engineering / Acoustics & Sound
Version 1.0, August 2022

TABLE OF CONTENTS

FOREWORD .... *I*
INTRODUCTION .... *V*

SECTION 1 \ CONCEPTS

CHAPTER 1: MINDSET – A CONSIDERED APPROACH 3
CHAPTER 2: GOAL-ORIENTED THINKING 7
CHAPTER 3: UNIFORMITY & VARIANCE 10

    Level Variance & Spectral Variance 11

    Complete Uniformity - A Worthy Goal? 11

    An Alternate Framing 15

SECTION 2 \ MECHANICS

CHAPTER 4: FREQUENCY, WAVELENGTH AND DIRECTIVITY 19

CHAPTER 5: TIME, PHASE AND SUMMATION 26
    Summation 27
    Effects of Level Offset 29

CHAPTER 6: OVERLAP 32

CHAPTER 7: BUILDING A LINE 36

SECTION 3 \ DESIGN

CHAPTER 8: THE SLIDING SCALE OF VARIANCE 47
    Range Ratio 47
    The Sliding Scale of Variance 48

CHAPTER 9: DESIGN CONSIDERATIONS AND WORKFLOW 55
    Point Source Aiming Considerations 55
    Line Array Aiming Considerations 57
    Subwoofers 65
    Fills and Delays 66

CHAPTER 10: FRIENDS IN LOW PLACES (CARDIOID SUBWOOFER STRATEGIES) 68
    Endfire 69
    Gradient In-Line 71
    Gradient Stack 72

SECTION 4 \ PERSPECTIVE

CHAPTER 11: THE MAGNITUDE FALLACY: PRIORITIES, TRADEOFFS AND FREE LUNCHES 77

Number of Affected Listeners..77

Main-Sub Alignment 79

Aux-Fed Subs...83

CHAPTER 12: MUCH ADO ABOUT MICROPHONES 88

Microphone Correction Files...88

Matched Microphones 91

Are All Mics Created Equal? 91

CHAPTER 13: SPLAY AND PRAY 94

0° Splay Angles 94

"That's not an array. THIS is an array!"......................................97

CHAPTER 14: THE CONFUSION ABOUT TILT 101

Uniformity is not Tonality 104

SECTION 5 \ ALIGNMENT

CHAPTER 15: PROCESSING AND DRIVE 109

Processing Granularity..110

Processing Structure 111

Three-Tiered Processing Approach...113

Signal Generator Routing...116

Generator Gain Structure 117

CHAPTER 16: THE FOUR PRINCIPLES OF ALIGNMENT 119

The Four Principles..121

Choosing Mic Positions...123

Array Zoning..125

High Frequency Shading 126

What About Gain Shading? 128

Fuzz at the Bottom 129

Multiple Mic Workflows 131

Ground Plane / Boundary Mic Positions 132

SECTION 6 \ STRATEGY

Chapter 17: Tuning Strategy 137

Studio L/R Nearfield Monitors 139

Coffee House Point Source L/R/Sub 141

Music Club L/R/Sub/Fill 144

Theater L/R/Sub/Fill/Underbalc 151

TV Studio (Distributed Mono) 155

Arena (Music Concert) 159

Arena (Installed System) 163

"Shed" with Lawn System 168

Chapter 18: Controlling the Variables 171

Defending the Show 177

Chapter 19: Common Pitfalls 179

Underspecified Systems 179

Underspecified Mains 180

Underspecified Side Hangs 181

Underspecified Subwoofers 182

Underspecified Front Fills 182

Front Fill Depth 183

Lack of Overshoot 184

Random Microphone Positions 184

0° Angles 185

Laser beam Centerlines 185

Relying on Autosplay 186

Skipping Verification 186

Filter Myopia 186

Tuning with Both Sides On 187

Prioritizing Gear over People 188

Conclusion 189

Recommended Reading 190

Further Resources 190

FOREWORD

There are many books devoted to audio engineering. And yet there are very few books focused specifically on live sound system engineering, design and optimization. In the modern era, virtually every system gets designed with prediction software and optimized with an FFT analyzer, making this a field well worth dedicated study. This book adds to that small number with a thoughtful study of the subject, presented in plain language with a plethora of practical information ready for immediate use by practicing system engineers, designers and mixers. It is a new book with a fresh, modern approach to the subject.

Audio is a very young form of engineering. Contrast this to the Stone Age roots of civil engineering or the 19th century electrical engineering of Tesla and Edison. Live sound engineering, in particular, is so young that many of its pioneers are still alive to tell the tales of its primordial era. The youth and rapid evolution combine to create a great challenge to those entering this field, since there are limited avenues of formal study to draw upon before venturing into the wild of on-the-job training or to supplement our skills along the way.

The 1st generation live sound engineers attempted to fill a technology vacuum that emerged when promoters booked rock bands into arenas and stadiums at a time when no sound reinforcement systems were capable of delivering a show that either the audience or the band could hear. My brother "saw" the Beatles in a baseball stadium in 1964, but he never claimed to have heard them. Nevertheless, these pioneers adapted to the challenge and developed tools that we might think have existed forever, like portable systems that could emerge from a truck and do a show on the same day.

In 1967, stage monitors and monitor mixing desks only existed if you built them yourself. Major rental companies made their own loudspeaker cabinets since leading loudspeaker manufacturers were too far behind the curve to make road-ready systems. The live sound engineer of that era needed a very different set of skills than those of the current day. Thiele/Small parameters, carpentry, passive crossover design, soldering, RC circuit analysis, transformer winding ratios, analog transistor and op-amp circuit analysis, grounding, matching amplifiers with appropriate drivers and most importantly putting up a ground-stacked PA so that it didn't tip over.

As a 2nd generation engineer, starting my touring in 1978, I was able to learn from the pioneers, but primarily through direct contact since they were still on the road. There also emerged a great number of important books, seminars and trainings wherein the knowledge gained by their experience was passed on. Many of these books had strong emphasis on physics and mathematics, most notably the incomparable works of Harry Olson. There is no arguing the value and validity of such deep explorations into our field of work, but there is also no arguing that one can be a great system engineer without understanding the difference between particle velocity and pressure waves. Such matters are required for the designers of loudspeakers and microphones, but not the users.

As my career moved on, I saw the skillset evolve and differentiate exponentially. Yes, there are still analog circuits, reflex ports to be designed and amplifiers to be matched with speakers, but these skills rest entirely within the scope of the manufacturers, not the everyday engineers doing shows. In the 1st generation, the people who built the gear were the same ones who used it, but this practice has become largely extinct as consoles, amplifiers, processing and loudspeaker systems evolved into the highly engineered tools of today. Two of the most significant advances of that period, in my estimation, were the emergence of FFT analyzers into the mainstream, and accurate acoustical modeling. These gave system designers and engineers self-sufficiency to predict the performance of their systems in advance of installation, and finely calibrate the system in place. These tools had the heavy math baked into them, allowing a new generation of engineers to focus on the decision-making process of system design and optimization with unprecedented levels of accurate information, all without a science degree.

A new approach is required as the current generation of live sound engineers moves to the forefront. This book clearly engenders the perspective of the modern engineer, of which Michael Lawrence is a card-carrying member. Michael has been a thought leader for his generation using his in-depth knowledge to both ask and answer the critical questions that emerge in our work. As a specialist in system design and optimization with real-world touring experience he brings all the required elements to the table in this book.
This book's emphasis is on practical decision making with the mature tools of the current day. You will not learn how to find the free air resonance of a cone driver. You will never need that to be a FOH, system or monitor engineer in the present or foreseeable future. Instead, you will find an orderly series of decision points presented in plain language, the very same ones we grapple with in venues each day.

There are very few black and white, clear-cut decisions in live sound reinforcement. Even polarity, seemingly the best candidate for such, is much grayer than you might think (as you will learn here). So many of the decisions we face in the design stage, such as array type, box count, aim, splay, height and more have a thousand possible answers. This book shows a logical process for the decision matrix involved to get to a final result or set of best options. Likewise, we face calibration decisions such as target curve, delay timing, fill levels and so much more, which have their own set of logical decision points. There is no single right answer to timing mains and subs, but there is a logical process to determining the best options. And no gig is complete without compromise, whether it be caused by budget, video screens, sightlines, lighting, weight or egomaniac architects and artists. Once again – decisions, decisions.

This book presents enough practical applications to age forward as systems continue their technological progression because it is based on rules of logic which will not go out of date. If reading this book makes you look back and reevaluate some of your past decisions, then that becomes a useful exercise for going forward, which is the only logical direction to take.

One last thing: you still need to learn to solder.

606 McCarthy
07/16/2022

INTRODUCTION

This is something I said I would never do.

The core of this book grew out of a series of brief and poorly conceived posts written for Reddit's r/livesound community, and were also endured by a small number of unfortunate individuals who happened to stumble across my website. In addition to my articles written for ProSoundWeb.com and Live Sound International magazine, my thoughts ended up sprinkled across the internet in a rather undignified manner.

As a growing number of friends and colleagues urged me to write a book (and one very dedicated Reddit user valiantly attempted to collect my output into an e-book of sorts), I started to wear down. I realized that writing a book didn't necessarily mean rehashing information that has already been communicated well by other sources (something I had no desire to do) but could be an avenue by which I could share my own views and approaches. The nature of this book became clear: a look from a different angle.

Despite the highly technical nature of the subject matter, this is not intended to be a technical reference on array theory, or the summation mechanism, or any of the core mathematical principles that dictate the behaviors and interactions of sound sources. Much like the forum posts that inspired it, this book aims to give a brief, conceptual explanation of these topics, just enough to give context to the discussions that follow.

For a more rigorous treatment, I strongly recommend purchasing (and reading) a copy of Bob "606" McCarthy's treatise on the topic, Sound Systems: Design and Optimization, which is widely considered to be the de facto foundational text on these topics. Like many others involved with sound system design and tuning work, I have benefited greatly from Bob's pioneering work that has informed that of countless audio professionals throughout the industry. I extend my most sincere gratitude to Bob for contributing a Foreword to this book.

This is also not intended to be a book on the use of audio analyzers, despite the pivotal role they play in a modern sound system tuning workflow. While I include some "tips and tricks" from my own personal workflow throughout the text, the reader is strongly encouraged to orient themselves with at least the basics of gathering and interpreting dual-channel FFT measurement data. Speed and fluency with an audio analyzer is the cornerstone of an effective modern sound system alignment workflow. In this regard, a significant portion of my professional efforts has been devoted to contributing to the numerous resources offered on this topic by Rational Acoustics, developer of the popular Smaart® audio analyzer, and all readers are encouraged to take advantage of them. (Users of other measurement platforms can rest assured that the measurement theory fundamentals still apply.)

This book is therefore best enjoyed as a companion to, not a replacement for, a study of Bob's book and a Smaart® training class. A list of additional recommended resources can be found at the conclusion of this book.

What you will find here is an explanation of my personal mindset and approach towards designing and tuning sound systems. It is a book about "how I think." I do not claim originality of anything contained herein. Rather the opposite: this is a largely derivative work, built off concepts and approaches learned, discussed, debated, and innovated alongside a large number of industry luminaries and colleagues, including Bob McCarthy, Pat Brown, Charlie Hughes, Merlijn Van Veen, Dennie Miller, Finlay Watt, Jim Yakabuski, Dave Rat, Carter Hassebroek, Fletcher McDermott, Tyler Walters, Daniel Lundberg, Bobby Brickman, Leo Pisaq, Kat Rice and Ville Kauhanen, as well as Jamie Anderson, Chris Tsanjoures and my entire family at Rational Acoustics, along with countless others, who are in turn building off the foundational work of early audio innovators such as Harry Olson, Don Davis and Richard Heyser.

Add this collective wisdom to a large pot, heat to a low simmer, add some salt, and once it cooks down and begins to thicken, you are left with my personal approach and mindset. (A special thanks to John Huntington for showing me how to make my book into a Real Book, and for introducing me to my fabulous layout designer, Shelbye Reese.)

This book is also - and this is important - not Gospel. What you are getting is one person's approach; a personal recipe, if you will. The reader is under no obligation to agree with my methods - indeed, I often find myself at odds with my past self. As uncomfortable as this can be, it is a clear indicator of progress.

Michael Lawrence
Fall 2022

SECTION 1
Concepts

CHAPTER 1
Mindset - A Considered Approach

We begin with a brainteaser of sorts. You are hired to design a sound system for a house of worship. When considering options for the front fill loudspeaker locations, you have the option of aiming them for either a seated or standing listening height. By observing the services, you know that the congregants spend about an equal amount of time seated and standing. Where, then, do you choose to aim the fills? Take a second and think through it.

It might seem like a 50/50 shot or simply a matter of preference. However, I would like to suggest that there is one option that is justifiably superior. I'll tell you which, and why - but not yet.

Designing and tuning a sound system is a long string of decisions and judgement calls. In my view, there are very few decisions to make along the way that are truly "fifty-fifty". In almost all cases, there is probably a justifiably superior option - we just need to apply some critical thinking and context to the decision, and with the help of a few "first principles", the way forward will become clear.

This, in so many words, is the core position I aim to advance within these pages. To condense this approach to a single sentence: have a reason for everything you do. As I have been known to tell folks I have mentored and trained, the goal is not to "do what I do." I would consider it a failure if my guidance produced a number of other audio professionals who were simply me-clones and always produced the exact results that I would. I'd rather they learn to focus on the thought process, and learn to consider context, variables, and tradeoffs in the way I try to. Nothing I do in the course of my work is random. Every product selection, splay angle, EQ filter, delay time, and microphone position is deliberate.

Building the habit of forcing yourself to justify your decisions – why you are doing something a certain way – is the fastest way to shine what can be a harsh light onto logical gaps and weak spots in your own approach. Perhaps someone taught you that way, or you were figuring it out as you went and never re-evaluated later.

This constant questioning leads to refinement and increased efficiency. And to me, this, more than any algorithm, product, tool, or technology, is the core necessity of successful sound system engineering work. This book is about the mindset and approach to the work which must drive the use of the tools, and be informed by the fundamental concepts.

For this reason, this book remains free of any discussion of specific loudspeaker products or technologies, apart from occasional mentions for illustrative purposes. I wish to extend my gratitude to Meyer Sound for allowing me to include screenshots from their prediction programs, **MAPP XT** and **MAPP 3D**, to better illustrate the concepts discussed within. The fundamental acoustic behaviors demonstrated by the predictions can be expected to occur with any comparable loudspeaker configuration, regardless of manufacturer.

Our focus here is on the underlying principles, which the reader may then apply to choose the right products for the task at hand.

Similarly, the measurement data screenshots are from Rational Acoustics Smaart® v8 audio analyzer, a tool with which I am intimately involved and deeply familiar. The data relevant for our purposes - magnitude, phase, coherence, Live IR, measurement delay - will be our guide through the alignment process, regardless of the measurement platform that is producing it.

And now we return to our house of worship front fill dilemma. A standing congregant is participating in their worship experience, whereas a seated congregant is an active listener to the words of the pastor. While the sermon is being imparted, the primary function of the sound system is to allow people to hear and understand the message, and during these parts of the service, speech intelligibility becomes paramount. Couple this conceptual reasoning with a more geometric one (when congregants stand, they are likely moving their ears further into the coverage pattern of the main system) and the way forward becomes clear - front fills should be aimed for a seated listening height.

Of course, there are all manner of circumstances that might make a stronger case for the opposite - and the "making the case" is the entire point. I care less about where your front fills end up, and more that you are able to justify why they're there. On the first day of tech rehearsal for a tour, the FOH mix engineer pointed to a particularly "interesting" looking EQ on my side hangs and asked how I had arrived at those filters. When I told him, he said, "that is the first time I've ever asked a systems engineer a question like that, and they actually had an answer." The considered approach is far rarer than you might think. The goal of this book is to make it a bit more common.

Imagine you have a high-spirited 5-year-old following you around as you work. They will ask "why" about everything – would you be able to answer them?

CHAPTER 2
Goal-Oriented Thinking

MIT electrical engineering professor Stephen D. Senturia once defined engineering as "the purposeful use of science." When we hit a rut, renewing our focus on the goal will guide our decision process and help us move forward. This can be a bit more challenging to mentally clarify for oneself in the heat of battle, but holding in your mind the purpose of whatever actions, choices, measurements, and decisions you are making at the moment is a powerful way to ensure that you are moving in the right direction.

From the metaphorical 10,000 feet, the goal is simple: end up with a system that puts sound where the listeners are, and not where they aren't. (Old "roadie wisdom" offers the sage advice to "point the loud side towards the audience.")

We always start by defining the audience area. Who needs to hear? Where are they located? A basic evaluation of the geometry of the audience area will yield some fruitful guidance on what loudspeakers we need, how many, where they should be placed, and where they should be pointed.

As we'll see shortly, it's more nuanced than this - hopefully we are setting our standards a little higher and striving for some amount of consistency - but the "first principle" here is that we will be given custody of an audio signal, and expected to transmit that signal faithfully to the listeners. To quote Merlijn Van Veen, we are the "waveform delivery service." As we work through our design and tuning - which I collectively refer to as sound system optimization, because there are a lot of things we can do to optimize (or not) a sound system at the design stage - we must always be cognizant of the design goals of the system, and make sure every decision made along the way works towards supporting those goals.

It is continually surprising to me how quickly this goal-oriented thinking can get lost, or become drowned in a soup of technical details. Perhaps the most common question during the monthly "Ask Us Anything" sessions offered by Rational Acoustics is "Where do I put my measurement microphone?" To which the response is "What are you trying to measure?"

Measurements answer questions, so we need to think about what we're asking, which will inform things like microphone position and analyzer configuration to help us answer the question.

If we want to know how heavy an object is, we should use a scale, not a ruler. If we want to learn about the level or tonality of a loudspeaker in its coverage area, the microphone should be located somewhere in that coverage area. If we want to learn about the timing relationship at the overlap seam between two loudspeakers, the microphone should be located at that seam.

What do you want to know about? Where should you place the microphone to investigate it? Often, application support situations and design dilemmas can be resolved by asking "What are you actually trying to do?"

Forgive me if this seems so straightforward that it draws into question whether it even deserves more than a sentence in passing. This is one of the repeat offenders when it comes to folks chasing their tail in tuning sessions, or developing "measurement myopia," staring into the analyzer and mucking about in the system processor without making any discernable improvements. I would like to be the little angel (or devil, I'm not picky), perched on the shoulder of the person tuning the system, tugging on their earlobe and asking, "What are you doing right now? And why?" If a considered answer is not forthcoming, I would suggest pausing the measurement activities until someone can explain what they are hoping to accomplish.

At every step, stop and ask yourself,

"What is my goal?"

CHAPTER 3
Uniformity and Variance

We turn now to the concept that dominates sound system optimization: the quest for uniformity. In the simplest of terms, we want a sound system that produces uniform coverage over the listening area. This doesn't mean "it sounds the same in every seat." This is impossible unless we're giving out a lot of headphones.

For a more practical definition, let's say that we want the sound of the show to translate well to all listeners, regardless of where they are in the audience area. The decisions made by the mix engineer about level and tonality carry little weight if the sound system itself creates a different level and tonality in every seat. Therefore, a consistent sound system that creates uniform coverage for the audience is what enables the mix engineer's creative decisions to be received by the listeners as intended.

As a rough analogy, you might think of how the color balance on a television must be set correctly in order for the visual language of a film to be properly displayed as the filmmaker intended.

In that sense, the value and necessity of a uniform sound system is self-evident whenever there is something that more than one person is trying to hear. This runs the gamut from simple speech intelligibility – a graduation ceremony or a political rally – to the subtle nuances of a musical theater mix, where attention is paid to the level and tonality at the single-syllable level. A uniform sound system makes that meaningful.

Level Variance and Spectral Variance

We can meaningfully characterize our progress in our quest for uniformity via the concept of variance – literally, how much does the coverage of the system vary over the space? Here we concern ourselves with two main types of variance – Level variance, which is the degree to which the system's coverage over the space varies in level, or colloquially, "how much louder is it in the front?", and spectral variance, which is the degree to which the system's coverage over the space varies in tonality.

Both level and spectral variance are easily measurable with an audio analyzer, and readily audible to a trained ear. They are both objective descriptors of the behavior of a sound system, and thus important parameters in any discussion about uniformity. A significant portion of Bob McCarthy's book is dedicated to a rigorous definition and exploration of these parameters. Here we focus on their conceptual and practical implications.

Complete Uniformity: A Worthy Goal?

Physics and geometry place some hard and fast limits on the extent to which variance can be eliminated from a sound system's behavior, and we will examine some of those in the upcoming pages. Conceptually, it's important to consider whether completely eliminating variance is actually our goal. In other words, if we have some magic technology that could make the system coverage exactly the same in every seat, should we use it?

My answer is "no," for several reasons. The first has to do with the context of the listener experience. Think about having front row tickets for Metallica or AC/DC, versus having lawn seats for the same event. The audience expectation for the environment is radically different. The folks on the lawn on picnic blankets would not fare well down front, being buffeted by the subwoofers and ducking under the crowd surfers. Conversely, it is my suspicion that the people expecting the visceral, immersive "front row experience" would find the lawn seats rather bland.

There is also an element of professional ethics here. Audience members self-select the type of sonic experience they want to have when attending an event. If you feel like taking a trip, visit my hometown of Rome NY for one of the weekly concerts held out on the City Hall green, and you'll see exactly what I mean: sixty feet of open space between the stage and line of elderly folks in lawn chairs. They know that it's going to be quieter in the back, and so that's where they want to sit. It's probably safe to say that most of these folks don't understand the intricacies of the inverse square law that governs the way sound sources lose level over distance. This is something they know in a practical, visceral sense, because it's how the world works. Too loud? Get further away.

If we design a sound system that has no front-to-back level drop, we have removed the ability for an individual to select for themselves a listening environment that is comfortable. Typically, this is not much of a concern in lower-SPL environments such as a Catholic church service or a Broadway musical, but as SPL rises, so does the practical need for a system with some level variance "baked in."

Psychologically speaking, it's important to understand the role of perceived proximity to the listening experience. Our brain uses a variety of psychoacoustic mechanisms to discern how far away a sound source is. The first - as just discussed - is level. When we get further from a source, it gets quieter, so level is a cue to proximity.

The second is direct to reverberant ratio. As we get further from a source, the direct sound falls in level, but reverb is (statistically) constant in level over the space. Up close to a source, the direct sound has an advantage, and as we move away, that advantage decreases as the direct sound must increasingly compete with reflections and general reverberation.

A well-designed sound system uses its directivity to focus the energy only into the listening area, and not onto the room's surfaces - in other words, designs work to maximize direct to reverberant ratio. We help it on its way with things like underbalcony fills - which often aren't about increasing level, but increasing intelligibility by adding in some fresh, high direct-to-reverberant-ratio energy.

A third cue is air absorption. Air is most absorptive of high frequencies, so distant sound sources seem "dull" to us. (Nearby thunder cracks, distant thunder rumbles.) The farther a loudspeaker must throw to arrive at its coverage area, the more severe a high frequency loss will be observed. Although a skilled system engineer takes this into account and will take steps to offset the effects, it might not be advisable to offset them completely.

A sound system that has low level variance and offers high direct to reverberant ratio has eliminated two of the cues that our brain uses to determine that the source is far away. Restoring the high frequency loss completely to the back rows of an arena can make a system sound uncomfortably close and create a strange psychological effect where the eyes are telling the brain that the source is hundreds of feet away, but it sounds as if it's RIGHT HERE (to get the full effect, please read that while holding your hand directly in front of your face). Thus, it probably makes more sense to decrease the system's spectral variance by partially compensating for air loss, rather than creating an unnatural and potentially unpleasant listening experience by eliminating it completely. Whether or not there is a giant image of the lead singer's face on a video screen might also be a factor when deciding how far to go.

My view on this whole concept of variance is that it is a design parameter: something we have the ability to control with intent, whether that means a system with a 3 dB drop front to back, or an 8 dB drop front to back, or whatever makes sense given the design goal (remember the goals bit?). It should not be something that we are forced into due to the inherent compromises made by the design. In virtually all cases, we will want to reduce the level variance over the listening area; the question becomes how much to reduce it.

Some quick back of the napkin math helps here. Calculating the Range Ratio of an audience area can help us understand what level variance we can expect if we do nothing to offset it. For example, consider a theater in which the closest seats are 16 feet from the stage, and the furthest seats are 160 feet from the stage. That 10:1 range ratio would result in 20log(160/16) or 20 dB of level difference from front to back. If it's loud enough to be comfortable in the rear, it's probably at a dangerously high level down in the front. If we want to reduce that to 6 dB level variance front to back, we will need to design a system that can nullify the remaining 14 dB of level offset. And now we have a place to start.

An Alternative Framing

Another way of thinking about a low-variance system is to imagine a sound system that creates the same coverage shape over frequency. As prediction software, measurement, or listening will confirm, a single full-range loudspeaker will have a narrower coverage pattern in the high frequency range than it will the low frequency range.

The frequency range over which the horn has custody will typically exhibit a controlled, directional response that can be aimed simply by pointing the loudspeaker in the desired direction. Not so for the low frequencies, which become increasingly omnidirectional as frequency falls. Stand behind a loudspeaker and listen – you hear the lows, not the highs.

Therefore, a loudspeaker has a varying coverage pattern over frequency, and as you move away from wherever it's axially aimed, high frequencies will decrease in level faster than the low frequencies, causing a tonal shift as you move off axis. Thus, since the low frequency energy has a wider dispersion than the high frequency, the loudspeaker has some spectral variance "built in."

Therefore, if we want a sound system that produces a consistent tonality throughout the listening area, we need it to produce the same coverage shape over frequency. We must either 1) widen the high frequency coverage shape, 2) narrow the low frequency coverage shape, or 3) some combination of the two. Effective system design will rely on both methods. The takeaway here is to be mindful of the fact that our sources will inherently create different coverage shapes over frequency, and consider that in our design and tuning decisions.

In a new venue, I go and sit in the seat furthest from the stage, and envision the audience experience from that seat. This offers healthy perspective throughout the design and tuning processes.

SECTION 2
Mechanics

CHAPTER 4
Frequency, Wavelength and Directivity

When trying to understand sound system behavior, a crucial thing to remember is that sound systems deal with a wide range of frequencies - if we establish as a generalization that our "full range" sound systems produce a frequency range from 30 Hz - 18 kHz, that's a 600:1 ratio in linear terms. Certainly, our brains perceive frequency in a largely logarithmic fashion, which makes this range about 9 octaves, but the mathematical reality that underpins the physical universe we occupy (sorry, you're stuck with it) unfolds in a linear fashion.

This has some extremely important consequences for system design and tuning. There are entire books dedicated to the behavior of sound waves, so of course you can gain a complete knowledge of the topic via the next three paragraphs:

Frequency (in Hertz) or "cycles per second," represented by an f, can be considered in terms of time as well. If we know cycles per second, we also know how long it takes to complete a cycle (**"Peri-**

od," represented by a T, or "Time Period" if you want to feel better about the T standing for something). These values are inversely related:

$$T = 1/f$$

The frequency of 500 Hz takes 2 milliseconds to complete a cycle. Since sound waves travel through air at a velocity of approximately 1125 feet per second (343 meters / second), we can also figure out how far the sound wave traveled in that amount of time. This is **wavelength**, represented by the Greek lambda (λ).
Frequency and wavelength are also related - divide the speed of sound by frequency to get wavelength, or divide the speed of sound by wavelength to get frequency. This can be described mathematically as

$$c = f\lambda$$

Based on this we, can calculate that 500 Hz has a wavelength of approximately 2 feet (0.6 meters).

Higher frequencies have shorter wavelengths, and lower frequencies have longer wavelengths. Imagine a car traveling down the highway at a constant speed, with a leak in the oil filter, dripping oil onto the road as it goes. Fast dripping (high frequency) means more closely spaced oil spots on the road (short wavelength). Slower dripping (low frequency) means more space between oil drops on the road (long wavelength).

There are a number of ramifications of the fact that our sound systems need to produce frequencies that span an enormous range of wavelengths (30+ feet at 30 Hz to under an inch at 18 kHz), but our focus here is on the relationship between wavelength and directivity. From the perspective of a high frequency (short wavelength), a loudspeaker's horn is physically large enough to direct the energy where we want it to go. This means that at high frequencies, aiming the sound energy is often as simple as pointing the loudspeaker in the desired direction.

This isn't the case at low frequencies, where the wavelengths are so long compared to the dimensions of a horn or a loudspeaker cabinet that the loudspeaker isn't able to exert much directional control over the radiating energy, and it becomes increasingly omnidirectional as frequency falls.

Sparing some more rigorous acoustical physics, this gives us a working conceptual explanation for why loudspeakers behave the way they do. Stand behind a "speaker on a stick," what do you hear? Predominantly low frequencies, because the highs are directed forward by the horn, but the lows radiate outward from the source more evenly in all directions.

By rough analogy, high frequencies radiate more like a flashlight beam (point it in the direction you want the energy to go) and the low frequencies behave more like the waves generated by dropping a stone into water.

Prediction software is the ultimate tool for visualizing and understanding this behavior, because it can show us the precise coverage pattern created by a loudspeaker over a particular frequency range. There are a number of loudspeaker prediction programs in common use, both manufacturer-specific and third party. The predictions in this book were produced by Meyer Sound's acoustical prediction programs, MAPP XT and MAPP 3D.

In prediction software, try adding a single full range loudspeaker of your choosing, and predict its behavior at 4 kHz (well into the frequency range over which the horn has custody and control) and again at 125 Hz.

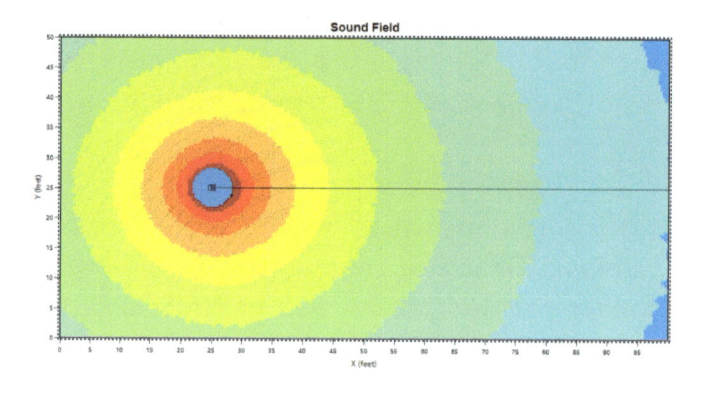

Figure 1 - Isobaric prediction of a point source loudspeaker at 125 Hz (MAPP XT)

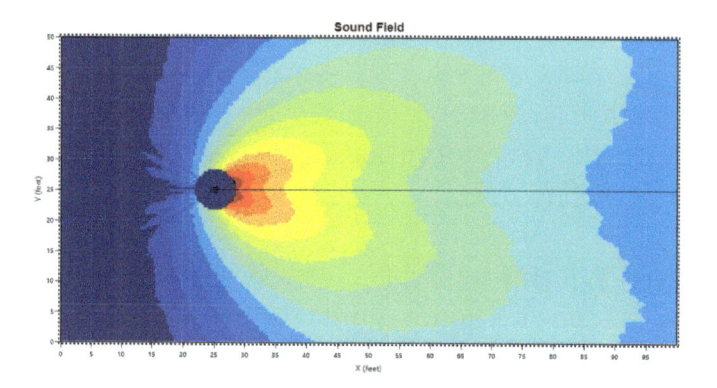

Figure 2 - Isobaric prediction of a point source loudspeaker at 4 kHz (MAPP XT)

These shapes look very different – this is the reality that forms our starting point for designing a system with controlled variance. If we want to design a system that makes a similar coverage shape at all frequencies, it's clear that we have some ways to go.

An immediate result of this reality is that we must turn a more skeptical eye to the specification of a loudspeaker's coverage angle. Consider the statement "According to the manufacturer's specification sheet, this loudspeaker has a horizontal coverage of 90°."

This specification is describing the coverage at high frequencies, where the horn has custody and the ability to exert some directional control over the energy leaving the loudspeaker.

The angle at which the output at high frequencies falls to 6 dB below its output directly on axis is considered the edge of the coverage. Start out front, move radially to the side, and if you made it 45° before the high frequency energy has decreased by 6 dB, you are the proud owner of a loudspeaker with 90° horizonal coverage at high frequencies. It might be much wider at 250 Hz.

This still leaves much information to be desired, of course, and not all loudspeaker designs are created equal. Most reputable loudspeaker manufacturers publish polar response data that describes the product's dispersion patterns at various frequencies, and of course prediction software will tell the tale as well.

As you might have guessed, there is an entire rabbit hole worth of information on this topic to be explored. For the purposes of this discussion, we leave that deep dive to more rigorous resources and content ourselves with the conceptual reality that the "90°" loudspeaker is much wider at lower frequencies.

Let's imagine placing two loudspeakers side by side, with their cabinets on an angle so the horns point in different directions. Perhaps we'll use two of the aforementioned 90° wide loudspeakers, and aim them 90° apart from each other, such that their off-axis edges meet at -45°. This would create a smooth, gapless coverage shape at high frequencies and everyone seated in the coverage area can hear. Wonderful.

If we splay them out wider, we develop a bit of a gap in the high frequency coverage. If we splay them out less, we have more overlap in the horn patterns, the consequences of which will be examined shortly. For the time being, we turn our attention to what's happening in the lower frequencies. The answer is overlap. Since both sources already had a low frequency dispersion pattern that's wider than the splay angle between the cabinets, we have some overlap happening.

Here we have the two fundamental mechanisms of how multiple sound sources interact with each other.

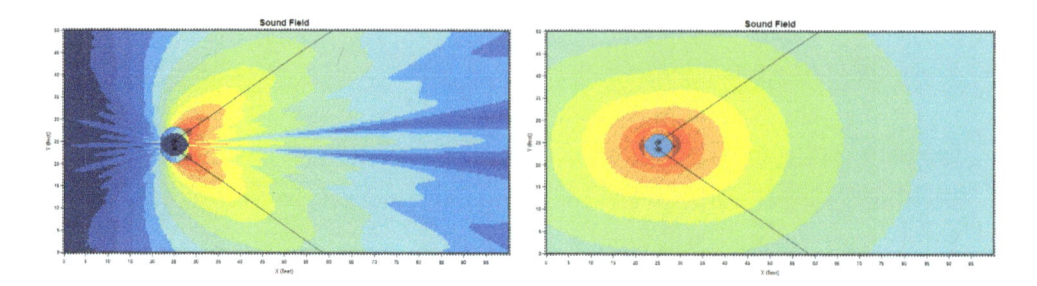

Figure 3 - Two 70° loudspeakers splayed by 70°, at 4 kHz (MAPP XT)

Figure 4 - The same configuration, viewed at 125 Hz (MAPP XT)

At high frequencies, any given listener is within the high frequency coverage from one of the loudspeakers, and adding a second one extends coverage angularly (another slice of pizza). At low frequencies, due to the overlap, *both* loudspeakers are contributing to the listening experience.

Thus, adding another loudspeaker adds high frequency *somewhere* and adds more low frequency *everywhere*. Extending the coverage radially with more loudspeakers also means a more low-heavy tonality from the combined sources overall.

24 \ *Chapter 4: Frequency, Wavelength, & Directivity*

Prediction software is your friend! Visualize the coverage pattern at different frequencies, and remember that the source is making all those shapes at once.

CHAPTER 5
Time, Phase, and Summation

Let's consider more closely the concept of Period (T), or the amount of time it takes a given frequency to complete a cycle. Since we're dealing in cyclical terms, we can refer to a full cycle as 360°, which allows us to describe partial cycles in terms of degrees rather than fractions. The trusty T = 1/f conversion is our friend here. It tells us, for example, that the frequency 100 Hz takes 10 milliseconds (ms) to complete a full cycle (360°). Similarly, a half cycle (180°) of 100 Hz happens in 5 ms, a quarter cycle (90°) happens in 2.5 ms, and so on.

Let's go down an octave to 50 Hz and consider those same time intervals - since a full cycle at 50 Hz takes twice as long (20 ms), we have only completed half a cycle (180°) in 10 ms, a quarter cycle (90°) in 5 ms, and an eighth of a cycle (45°) in 2.5 ms.

We use the term **phase** to describe where a certain frequency is in its cycle, in degrees. Higher frequencies will cycle through more degrees in a given amount of time. As an analogy, think about placing a bunch of tires of different diameters on the starting line at a racetrack, and roll them all forward at the same speed. They'll all cover the same distance, but the smaller tires will make more revolutions along the way. This means that, in a given amount of time, different frequencies go through different portions of their cycles. 1 millisecond is equivalent to 36° at 100 Hz, 180° at 500 Hz, 360° at 1 kHz, and 720° (two full cycles) at 2 kHz.

Summation

This all seems a bit academic until we start thinking about how multiple signals interact. Imagine we are mixing a signal with a copy of itself that is equal in level but delayed in time. You can easily create such a situation in Digital Audio Workstation (DAW) software by copying a track and sliding the waveform slightly. For simplicity, let's imagine the file contains a sine wave (all the energy concentrated around a single frequency). The amount of time offset between the signals dictates phase offset for a given frequency, and it's the phase offset that determines the nature of the resulting summation. At 0° offset, the two waveforms are completely aligned, resulting in a textbook-perfect summation – a doubling in level (+6 dB). Introducing some time offset will start to degrade the quality of the summation, but the sum will still be greater than the level of the individual contributors as long as the phase offset remains within 120°. Once we hit 120° of phase offset between our signals, we've lost the benefit of summation, and the resulting sum is the same level as the original contributors.

Past 120° of phase offset, the two signals start to fight each other – peaks increasingly align with valleys – and the resulting level is lower than the original contributors. 180° is the "worst case scenario", where the maximum positive peak in one copy of the signal is perfectly aligned with the maximum negative peak in the other, and they counteract each other to result in cancellation (a theoretical negative infinity, no level).

You can calculate the resulting summation for any phase offset, but it is practical and helpful to commit a few milestones to memory. Assuming the two contributors are matched in level:

$$0° \text{ phase offset} = +6 \text{ dB}$$
$$90° \text{ phase offset} = +3 \text{ dB}$$
$$120° \text{ phase offset} = 0 \text{ dB}$$
$$150° \text{ phase offset} = -3 \text{ dB}$$
$$180° \text{ phase offset} = -\text{inf dB}$$

Now we come to the fun part: a given time offset will result in different phase offsets for different frequencies. That means when we mix our signal with a delayed copy of itself, that time offset creates a summation that increases level at some frequencies and decreases level at others. Frequencies at which the two arrivals fall within a third of a cycle (+/- 120°) will see addition, the rest will see subtraction. The resulting response is an alternating series of peaks and dips over frequency.

This pattern is called a **comb filter** because it looks like the teeth of a comb when viewed on a linear frequency plot.

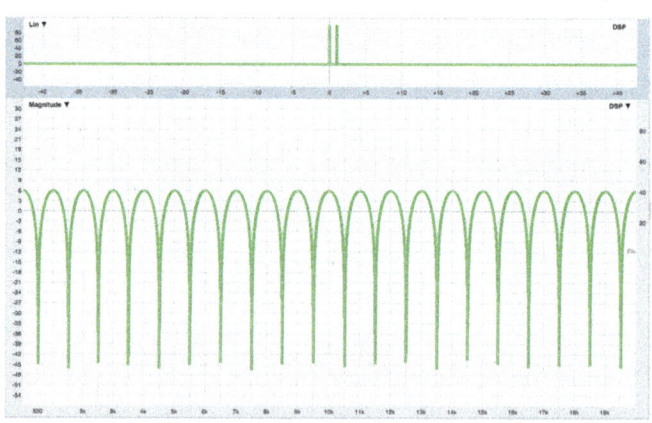

Figure 5 - A 1 millisecond comb filter. The top pane (Impulse Response) shows the two arrivals separated by 1 ms in the time domain. The bottom pane shows the resulting comb filter response on a linear frequency axis. (Smaart® v8)

On a logarithmic frequency axis, which more closely resembles our tonal perception, we see closer spacing of peaks and dips as frequency increases. Increasing time offset extends the comb filter to lower frequencies and pushes all the peaks and dips closer together.

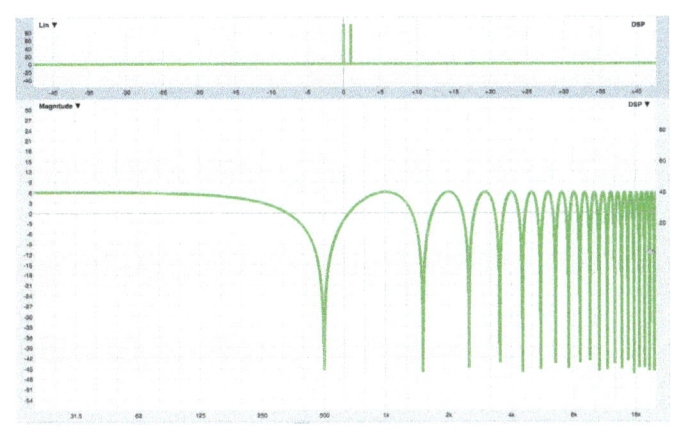

Figure 6 - The same 1 ms comb filter, displayed on a logarithmic frequency axis. (Smaart® v8)

Effects of Level Offset

The nature of these interactions will hold true regardless of any level offset between the sources, but the addition and subtraction are most substantial when the two sources are matched in level. As level offset increases, the higher-level source increasingly dominates the summation. The height of the peaks and depth of the dips both decrease, because the lower-level signal has less influence in the summation. Whenever one source has a significant level advantage (let's draw a line in the sand around 10 dB), the contributions of the lower-level signal become far less significant.

BETWEEN THE LINES / 29

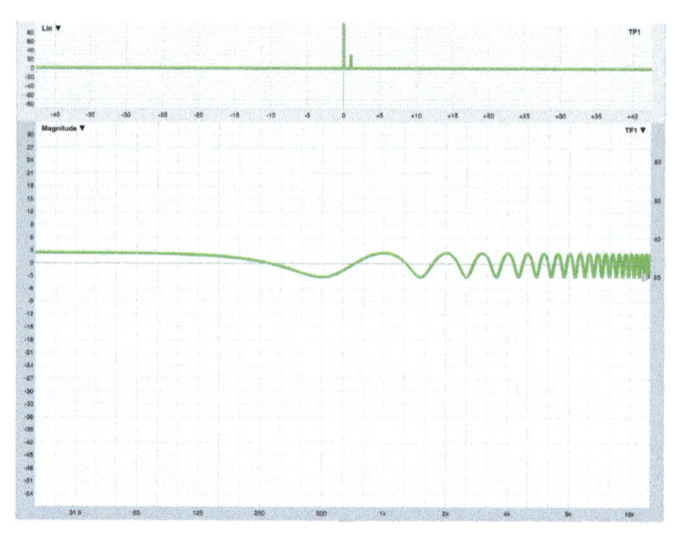

Figure 7 - The same 1 ms comb filter, with the second arrival attenuated by 10 dB. (Smaart® v8)

This leads us to an important point that will become one of our "first principles" for our design and alignment approach: **sources are most interactive where they are equal in level.** When there is a significant level offset, the higher-level source wins. When levels are similar, the phase relationship determines the outcome.

Now that we see the effect that time offset has in the frequency domain, let's consider how that affects the real-world behavior of the loudspeakers within our sound systems.

"Are these two sources in overlap, or isolation?" The answer might be "Yes!" because of different directivity at different frequencies.

CHAPTER 6
Overlap

This chapter explores one of the most critically important concepts in sound system design: how overlapped sources work together (or don't). Whenever we have multiple sources reproducing the same signal into the same coverage area, we have an opportunity to use their combined output to increase the level in that area, which is a powerful tool for designing systems that can achieve the required sound pressure level (SPL) for an application. If a single loudspeaker can't produce the required output, the options are either get a more powerful loudspeaker or use multiple loudspeakers working together. Very generally speaking, we use the former solution at high frequencies and the latter solution at low frequencies. Let's explore why.

To begin, we return to the example configuration we had previously: two loudspeakers placed side by side and splayed so their high frequency coverage shapes align at their off-axis edges. This extended the high frequency coverage horizontally, and gave us increased level at low frequencies.

At high frequencies, the directivity of the horn means that the two sources operate in isolation – depending on where you stand, you're either in the coverage of one horn, or the other, with a small overlap at the seam between the two.

Low frequencies, however, have a different reality. Regardless of where you stand as a listener, you will get energy from both sources at low frequencies due to their lack of directivity.

The question we need to answer is "How do the two sources interact?" Relative arrival time is the key to this puzzle. The same signal is sent to both loudspeakers, and they both reproduce it at the same time. If we are standing in a location that is equidistant from both sources (out in front, for example), arrivals from both sources will reach our ears at the same time. As soon as we take a step off to the side, we're closer to one source and farther from the other. They now have unmatched arrival times at our listening position. It's this time offset between the arrivals that governs the resulting response at that position.

Let's put some numbers to it. Two sources with no time offset (equally matched in time) will sum together perfectly and create the maximum possible summation (+6 dB for level-matched sources). As we move off center, one source arrives before the other to our listening position, and this is the situation we've been mentally preparing for: two instances of the same signal with a time offset between them.

At low frequencies, the time offsets we accrue by moving off-center from our pair of side-by-side loudspeakers produce inconsequential amounts of phase offset. For example, 100 Hz has a period of 10 ms and a wavelength of about 11 feet. If we move off to the side until one source is 6 inches closer to us, that small path length difference will cause a time offset of approximately 0.5 ms, which corresponds to a twentieth of a cycle (18° phase offset) at 100 Hz – well within the summation zone of <120° phase offset.

At high frequencies, we don't have nearly as much leeway. That six-inch path length is quite substantial indeed from the point of view of 4 kHz, which has a wavelength of about 3 inches, so the offset corresponds to 720°, two full cycles. Around 666 Hz, the scales tip past 120° phase offset, and frequencies above 666 Hz have descended into ~~the depths of hell~~ combing.

This is why we aimed our loudspeakers so the horns only met at the off-axis edges of their coverage pattern, rather than having more overlap. The spatial offset between the two horn mouths is enough to cause a comb filter at high frequencies, so we need to make sure to use angular isolation to prevent them from overlapping. At low frequencies, however, the long wavelengths make the physical displacement between the loudspeakers rather minor by comparison, and the resulting phase offsets are small enough that the two sources side by side give us some useful low frequency summation.

As the physical displacement between the sources increases, we will experience more variation in the relative arrival times as we move around the listening area, and thus the response will be more varied over the space. The takeaway here is that physical displacement is a critical factor in the resulting response when sources have overlapping coverage, which means our job is to carefully manage overlap by minimizing physical displacement between the sources. At high frequencies, the wavelengths are shorter, and our 120° countdown clock runs out more quickly, so we use the directivity of the horns to carefully constrain energy from each high frequency source to its own coverage area.

Looking ahead, our alignment process will hinge upon paying

careful attention to these seams where our sources overlap at high frequency. At these locations, our two sources meet at equal level, and so the summation is most sensitive to time offset, so we'll manage it carefully.

> *Two sources would seem a lot farther apart to an ant than they would to an elephant. When considering source positions, we must consider the displacement from the perspective of both low and high frequencies.*

CHAPTER 7
Building A Line

Audio engineers really seem to enjoy making lines of objects, especially faders, road cases, and cable runs. In this chapter we'll take advantage of that psychological urging and examine what happens when we make lines out of our sources.

For a look at the fundamental mechanism, let's begin with the most omnidirectional of audio sources: subwoofers. Subwoofers are effectively (although not perfectly) omnidirectional throughout their operational bandwidth, let's say 100 Hz and below. That makes them useful for studying the effects of overlapped coverage in a line source, since multiple subwoofers are effectively 100% overlapped at all locations.

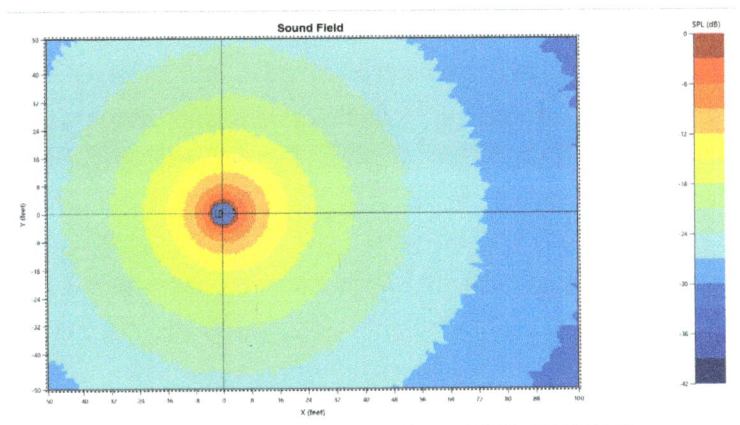

Figure 8 - A single subwoofer at 50 Hz (MAPP XT)

Perhaps counterintuitively, it is this overlap that allows us to arrange multiple subwoofers to create directivity. This is important for its own sake – unlike our full-range loudspeakers, we can't focus the subwoofer's energy in the desired direction simply by pointing it where we want it to go – but also because this mechanism is the primary (and often misunderstood) force behind the behavior of a modern line array at lower frequencies. Let's understand how it works so we can figure out how to make it do what we want.

We start by assembling six subwoofers into a line and observing the effects.

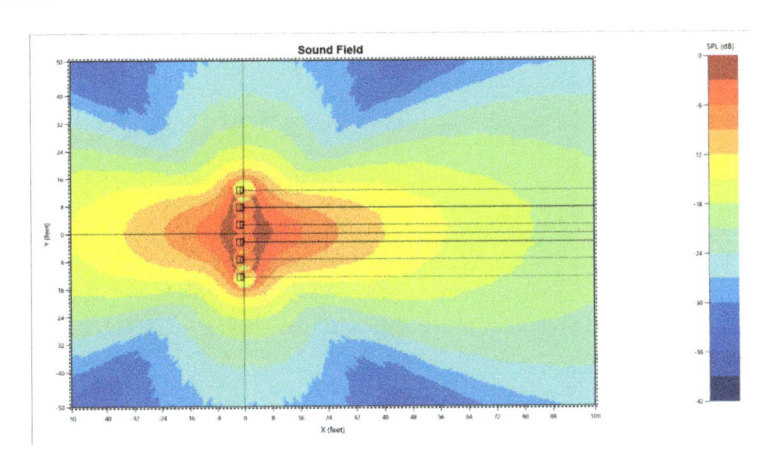

Figure 9 - Six subwoofers in a line 25 feet long, shown at 50 Hz (MAPP XT)

BETWEEN THE LINES /37

We can see that the coverage shape is no longer a nearly-omnidirectional blob, and has instead formed a narrower and more focused blob, although to avoid hurting its feelings we'll refer to it as a "lobe" when it is within earshot. What accounts for the narrowing of dispersion?

Simply put, this is phase offset at work. In front of the array, the energy from all six sources arrives within quick succession (small phase offset), and so sums together constructively. On the sides, however, each element is a different distance from the listening location. All the arrivals are staggered in time (larger phase offset), so we don't get the same coherent summation that we enjoyed out front. The underlying mechanism here is that time offset between arrivals is smallest in front of the array (most summation) and largest on the sides of the array (least summation).

Production management is feeling generous and has approved our request for more subwoofers. Let's add more, but pack them more closely so that we *keep the overall line length the same.*

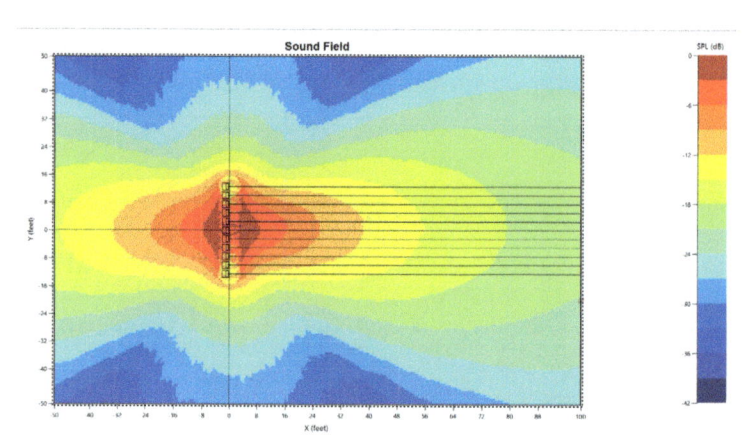

Figure 10 - Same line length, but more elements spaced more closely. Coverage angle is still the same (50 Hz) (MAPP XT).

Now this is interesting –we have more output (more sources = higher

SPL) but the overall dispersion angle of the blob, er, lobe, is unchanged. This is because the beam narrowing mechanism is driven by the length of the line. If we now keep the same number of elements, but space them out more, we can further narrow the dispersion by creating more phase offset on the sides of the array.

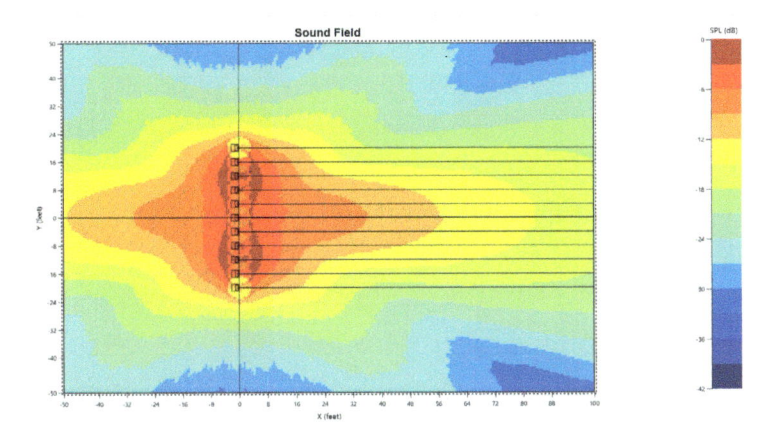

Figure 11 - Same number of elements, longer line, narrower coverage angle (50 Hz). (MAPP XT)

We have now learned that, throughout the frequency range over which the individual elements are sufficiently overlapped, increasing line length narrows the coverage in that plane. We must remember that displacement (physical distance between boxes) is different from the perspective of different frequencies. A 16 foot long line is one wavelength long from the perspective of 63 Hz, and two wavelengths long from the perspective of 125 Hz. For this reason, we should expect the coverage pattern to be twice as narrow at 125 Hz as it is at 63 Hz.

The converse is also true: a 16 foot long line will produce the same coverage angle at 63 Hz that a 32 foot long line will produce at 32 Hz. In both cases, the line is one wavelength long. This reminds us of one of the fundamental challenges of sound system design: our sources make different coverage patterns at different frequencies.

Luckily we've just stumbled upon a mechanism to get some control over dispersion at low frequencies.

High frequencies steer with magnitude ("point the loud side towards the audience") and low frequencies steer via phase (the relative time offsets and resulting phase offsets create beam narrowing). The difference is overlap. A very short line length, or a single element, can't create much directionality at low frequencies. A physically longer array can exert more directivity to lower frequencies, and so line length becomes an important design parameter. This is what some sound engineers are referring to when they use the term "control."

The line of subwoofers has the most output directly on-axis with the geometric center of the array because that's where we have the closest to "equal time, equal level" meeting point for all the sources, and is sometimes termed "power alley" by mix engineers. If we want to change the directivity of the beam, we can either physically move the subwoofers (bend the line into an arc, for example) or use electronic delay processing to virtually bend the line into an arc.

Let's lift our sub array up from one end and hang it from the ceiling as a vertical line. Now we have an array that narrows coverage in the vertical plane instead – less energy hitting the ceiling and floor, and more energy forward into the seating area of the venue.

This is the same basic behavior we can expect out of a full-range line array at these lower frequencies. At high frequencies, each element has a very narrow vertical coverage pattern, and we can point it where we want it to go. The top elements need to go farther to reach their intended listeners, so we use smaller angles, which means more overlap and more summation in the high frequency range. Towards the bottom of the array, listeners are closer, and we open the splays wider for less summation over the shorter throw distance.

Regardless of the chosen splay angles, the individual elements have a much wider pattern at lower frequencies, and are therefore highly overlapped, letting line length take over to create the directivity. At high frequencies, a line array creates its coverage shape via magnitude (isolation afforded by the highly directional horns) and at low frequencies, the coverage shape is created via phase (line length creating time offset due to the overlapped elements).

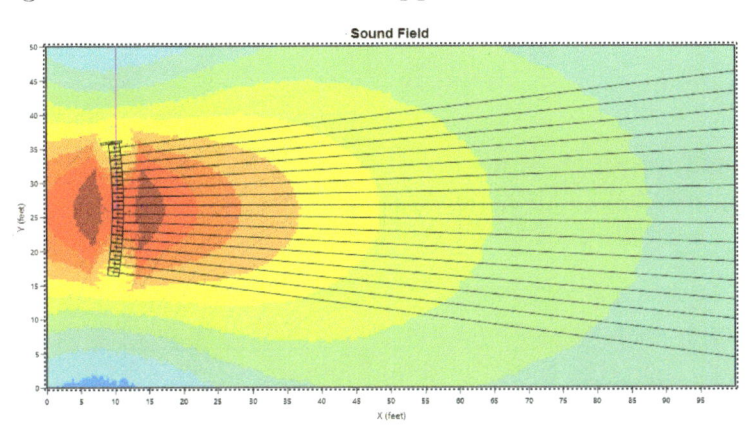

Figure 12 - A 16-element flown array viewed from the side (Section View) at 50 Hz (MAPP XT)

There's a bit more to it than that, but this is a good functional starting point for understanding how these arrays work. By extension, this means that when you're sitting in the coverage of a line array, you are listening to a handful of elements at high frequency (the ones whose horns are aimed at you) and all the elements at low frequency.

This has important ramifications for our tuning process. Since all the sources contribute to the coverage shape at low frequency, we won't find success by trying to EQ the low frequencies out of a single box. If we end up with an excess of low frequency energy in the front of the coverage area, we can't simply turn down the low frequencies in the bottom few boxes, because the entire array is contributing.

Conversely, if we have too much high frequency energy down front, it's very easy to simply apply some high frequency EQ to the few elements that sound bright, without concerning ourselves with disrupting what's happening elsewhere – because "elsewhere" is covered by different elements at high frequency. Meanwhile the low and low-mid beam shape will always emanate from the geometric center of the array. This might not always be a good match for the shape we're making with the higher frequencies, which tends to be asymmetrical with more output towards the top).

You might wonder how a line array-type element gets away with overlapping coverage at high frequencies without creating objectionable comb filtering.

The key to success here is twofold: first, high frequency waveguides that are designed to keep the acoustic sources in close physical proximity, which reduces the relative time offset and therefore phase offset as we move off axis in the vertical plane. The equation is completed by the very narrow vertical dispersion of the horns. Narrow coverage in the vertical plane means that we fall out of the coverage of the horn before we get far enough above or below to accrue enough arrival time offset between the adjacent boxes to cause a severe comb filter.

That's not to say that we avoid the issue altogether, and keeping line source elements in coupling behavior above 8 kHz or so can be a tall order due to the tiny wavelengths involved. However, a lot of manufacturer R&D efforts are dedicated to creating a configuration that keeps the sources behaving well as a group. Whereas if we "home build" an array by stacking up multiple point sources with wide coverage, we can get far enough off axis to encounter enough time offset to comb filter at high frequencies, while still being within the coverage of the horns, and that's where the problems start. The narrow vertical coverage of line array elements minimizes the time offset that it's possible to accrue while still in the coverage of multiple elements.

Line length is a design parameter, not just a logistical issue.

SECTION 3
Design

CHAPTER 8
The Sliding Scale of Variance

We begin our foray into my system design philosophy with the acknowledgement that the audience area is the boss. The shape and size of the area occupied by the listeners informs decisions as to what equipment we need and where to point it.

Range Ratio

A good starting point is an evaluation of Range Ratio, which is the ratio of the throw distances from the sound system to the nearest and most distant listeners. We'll then use that information to design a sound system that can create appropriately asymmetrical coverage. For example, if we are asked to design a system for a venue where the closest listeners are 18 feet from the stage and the furthest are 90 feet, that's a 5:1 ratio, or $20\log(5/1) = 14$ dB between front and rear if we do nothing to reduce the level offset. This is our good old friend the inverse square law at work. The higher the range ratio, the less likely we are to be able to achieve a low-variance system by basic means (a pair of loudspeakers on sticks, for example) and the more likely we are to benefit from more advanced tools such as flown arrays and delays.

As we discussed at the beginning of this book, our goal is not necessarily to completely eliminate the front-to-back level offset. Like cheesecake and Cheetos, "louder in front" is often a desirable thing when enjoyed in moderation. If our design goal is 6 dB of level variance from front to back, then we need a system that can counteract the inherent 14 dB range ratio by 8 dB. Having an idea of where we are starting and where we are trying to get is important context for the decisions that follow.

The Sliding Scale of Variance

Although our aim is to reduce both level and tonal variance, these are often – but not always – opposing goals. By way of example, let's consider a local theater I work in a few times a year. The vendor supplies 6 boxes per side of a small-format array product. The figure below shows the section-view (side view) predictions at 200 Hz and 4 kHz.

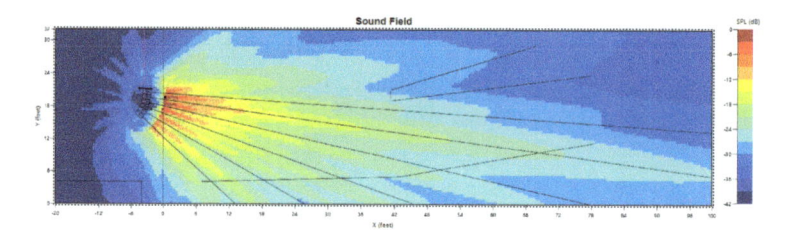

Figure 13 - Six-box hang at 4 kHz (MAPP XT)
The splay angles create even coverage from front to back at high frequency, but at 200 Hz, we're not able to keep up thanks to the relatively short line length (just under 6 feet).

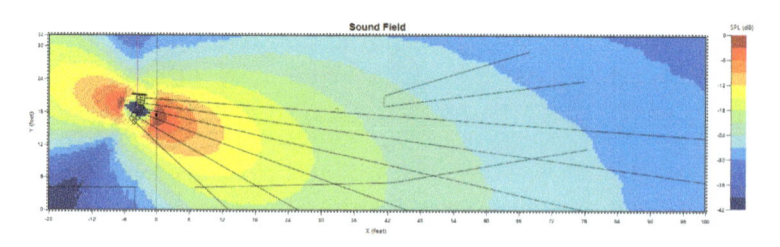

Figure 14 – The same six-box hang at 200 Hz (MAPP XT)

48 \ *Chapter 8: The Sliding Scale of Variance*

Let's pause to examine the coverage shape at 200 Hz, where wavelength is approximately equal to line length. Note the Figure-8 shape. This is an important milestone to familiarize ourselves with. Above this frequency, line length exceeds wavelength and the array begins to develop increased directivity. Below this frequency, dispersion becomes increasingly omnidirectional. It's important to understand that this is a gradual progression - taking a single box off the array doesn't instantly create omnidirectional mayhem (although that's a great name for a band) - but this is a good milestone.

Recall the mechanism by which line length narrows coverage: staggered arrival times above and below the array. The figure 8 shape emerges because at this frequency, the staggered arrivals above and below are as staggered as they possibly can be - they run a full 360° cycle (because the line is one wavelength long) and so we see the maximum "scrambling" of arrival times, and maximum cancellation. If we were to double the line length to 12 boxes, we would be able to push the Figure 8 shape down an octave to 100 Hz, and gain some additional directivity at 200 Hz.

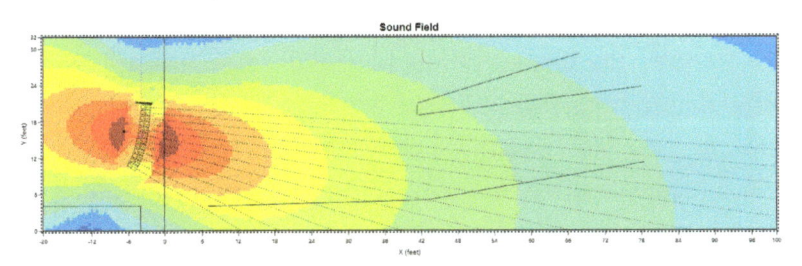

Figure 15 - Doubling the line length now creates the Figure 8 shape an octave down (100 Hz). (MAPP XT)

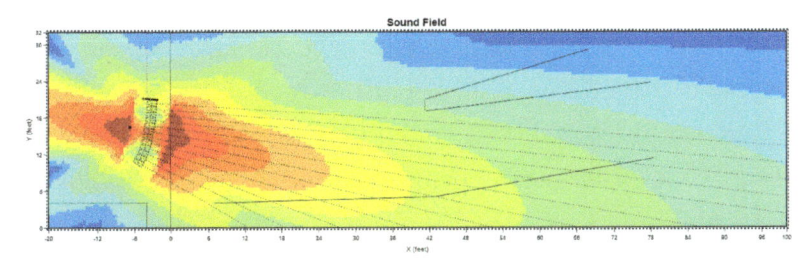

Figure 16 - Meanwhile at 200 Hz we have gained additional directivity (MAPP XT)

Given the budgetary constraints, and with the realization that the line length dictates the array's potential front to back uniformity in the low frequencies, it would be inadvisable to tune the system so the high frequency is perfectly consistent from front to back, because this would lead to a system that sounds too bright in the rear, or too bass-heavy in the front. In other words, a system that has perfectly consistent coverage at high frequencies, but more low frequency energy towards the front, will exhibit tonal variance front to back.

The alternative approach would be to let the high frequency taper off from front to back as well – keeping the tonality of the system consistent over the space, at the price of some level variance (louder in the front). The tonality of the mix will now translate well, but if we assume the FOH mixer is running the show at a comfortable level from their mix position, it could be uncomfortably loud down front.

We can imagine our variance goals as a gradient or sliding scale, with low spectral variance at one end and low level variance at the other. A tuning strategy that reduces spectral variance likely increases level variance, because the line length dictates the variance in the low frequency region regardless of what we do at higher frequencies. If we can increase line length, we can move the two ends of the variance scale closer together, and come closer to achieving both goals at once. Increasing from 6 boxes to 12 goes a long way towards creating a more focused dispersion at low frequency and puts us in the realm of 6 dB drop from front to back at 250 Hz.

We don't need the additional boxes to make the show *louder*, as we may be able to achieve our required SPL (whatever that may be) with just 6 elements per side. The overall uniformity, however, is greatly improved by the additional loudspeakers, and the extended line length is a major part of that equation. There are also some benefits in terms of high frequency uniformity that we'll explore shortly.

For this reason, choosing the minimum number of elements required to achieve target SPL or simply cover the required vertical coverage angle is often a recipe for a system that has very high variance, and adding a few more elements is an easy way to take meaningful steps towards lower variance.

In this sense, many systems are under-specified. It's easy for a production manager or account rep to look at the increased line length and say "you could do this show with half the PA." From a standpoint of achieving the required SPL at mix position, that might be true. However, arrays consisting of more cabinets and longer lines are very important when our goal is not just SPL but also reasonably low variance. For this reason, I typically start my design process with the maximum number of elements possible, considering availability, weight, and sightline restrictions, and work my way down. "How many boxes can I get?"

Of course, it is possible to have too much of a good thing. Particularly with low trim heights, longer lines collapse the splay angles because we have more elements covering the same vertical coverage angle. If we tried to hang 16 boxes in this theater, for example, we end up with a line full of 1° angles, which is unhelpful since we're trying to use a progression of splay angles to reduce front to back variance at high frequencies.

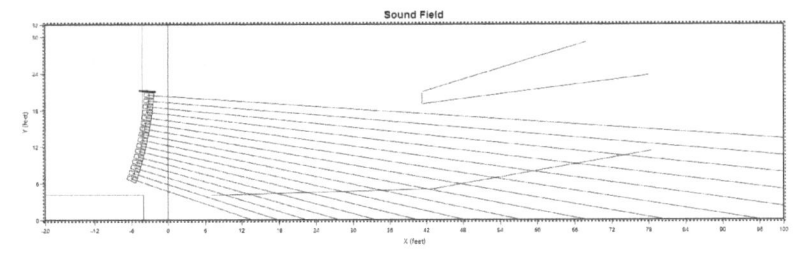

Figure 17 - This has become silly. (MAPP XT)

Sight lines become a concern, as does weight load on motors and rigging points (as does having the bottom box of the array firing directly into an audience member's face).

In situations with larger vertical coverage angles, a longer line can conceivably concentrate the low-mid beam *too much* until it's narrower than the coverage at high frequencies. This sets the table for a very unpleasant listening situation for those seated near the top or bottom of the array's coverage. Recent years have seen manufacturers paying more attention to proprietary electronic processing techniques to steer / spread the low-mid beam. As a general rule, however, it is best to get as close as possible to our coverage goals with mechanical means, and let the processing carry us over the finish line.

And let's not forget about the subwoofers. Often in small and medium scale applications, flying subwoofers is not logistically reasonable. We're stuck placing them on the ground, resulting in the subwoofers being much louder towards the front of the listening area.

Even though we may succeed in designing a wonderfully behaved main hang that accomplishes very low level and tonal variance, the subs are still going to create excess SPL towards the front, so we need to keep that fact in mind as we make decisions about the variance of our full-range systems.

We do have a couple of factors working in our favor. The first goes back to our discussion about audience expectations. Folks sitting in the front row of a music concert *expect* more subwoofer energy for many music genres, Acoustic Terry notwithstanding. The front row audience members' expectation for a more sub-heavy experience conveniently align with the fact that our systems will naturally have more sub energy down there. This means some portion of the increase in sub tilt towards the front of the room is a "good thing" and indeed a design parameter for many types of program material. (In designs with flown subwoofers, it is common to place a number of subwoofers on the ground particularly to offer control over the sub energy in the front rows during the tuning process.)

The other thing we have going for us is that the system variance is easiest to control throughout much of the vocal range, and for a lot of events, variance in the vocal range is arguably the priority, so the situation is often not as dire as it might seem.

In all cases, we must be mindful of variance over the entire frequency spectrum. It's very easy to become hyper-focused on the 1 kHz – 4 kHz region, which is where our splay angles and high frequency equalization are very effective, and end up making that region extremely consistent without paying any mind to what's happening in the rest of the spectrum. This won't result in a consistent-sounding system if the program material is full bandwidth.

By the same token, it's easy to open prediction software and adjust the model until the entire listening area is the same color at 4 kHz, even though in practice this will result in a system that is tonally inconsistent because it doesn't consider the entire spectrum.

Since the geometry of the system design is what sets the baseline for the system variance at lower frequencies (in this case, ground-based subwoofers and short line length), I like to start by gaining an understanding of what the system is doing at those frequencies, which informs my decision of how far I want to push for uniformity in the highs, keeping in mind the particular use case of the system and the program material. Some applications might lead me to prioritize spectral uniformity over level uniformity, or the other way around.

A prime example would be the same system in use for two different shows: a loud rock band versus a solo performance by local favorite Acoustic Terry. For the solo acoustic performance with negligible sound level coming off the stage, I would probably tend to keep the level ramp-up towards the front as minimal as possible, since I don't need to worry about them being overwhelmed by stage wash, and there's not a lot of sub-bass information that will make the sub ramp-up undesirable. In other words, prioritize level variance at the expense of some tonal variance.

For the full rock band, the acoustic energy of the drums and bass and guitar amps coming off the stage means that it is probably desirable to let the bottom zone of the array and the front fills gain a few dB of level, which keeps the overall spectral variance lower, at the expense of some level variance. The system is louder towards the front of the listening area, but is more spectrally consistent because the full-range mains are also higher in level towards the front, not just the subwoofers. The tradeoff between spectral and level variance, particularly at the front and rear "edges" of the system (front fills and lawn system / delays) is a call that I will make on a per-event basis after consideration of all the variables involved.

What are the artist's expectations for how the system sounds and feels in the front row?

CHAPTER 9
Design Considerations and Workflow

My approach to design follows the same general pattern that the tuning process will follow once the system is deployed: start with the biggest components that cover the largest share of the audience listening area (mains). Work to get the mains producing consistency within their custody area, then figure out how big of a coverage gap we still have, design a sub-system to fill that, and repeat the process.

Along the way, we will pay close attention to the seams between the different sources, and try to design our system in a way that enables us to make meaningful and effective timing decisions at these seams. This chapter is not a full system design course (if you've been following along, you know where to find that) but offers general insights into my personal approach.

Point Source Aiming Considerations

We start by defining the custody area for our mains (how many seats have been sold? Where are the listeners located?) Then we have two decisions to make regarding aim: horizontal and vertical. Horizontally, find the center of the custody area and aim through it. If you're covering a seating section that is 15 seats wide and 9 rows deep, aim through the 8th seat in the 5th row. Center of center. In virtually all seating geometry situations, this will be the best compromise of coverage. If it doesn't cover all the seats, you either need a wider source, or an additional one.

In the vertical plane, our options for offsetting range ratio with a point source are limited, but we can get the most benefit by aiming axially directly at the farthest listening position we need to cover with this source, because that's where we need the most output. From there, moving closer to the source also moves us off-axis (provided, of course that the source is placed above head height) so we offset the proximity gain with the axial loss, at least to an extent.

Once we get to the -6 dB bottom edge of the coverage, we've run out of horn and we need to call in reinforcements in the form of another loudspeaker if we've still got more ground (ears) to cover.

If you've got a 'speaker on a stick' situation, the stand height might be the only variable you have control over. If our loudspeakers are placed directly on the stage deck, we turn to the time-tested "acoustic aiming device" a.k.a piece of 2x4 to get the right tilt to aim our mains at the last row.

A dose of realism here: particularly with the smaller 'speakers on sticks' variety of gigs, we often "get what we get" in terms of loudspeakers, and simply must do the best we can with what we have, rather than being able to select our choice and quantity of loudspeakers from an inventory.

What we do in such situations is simply *minimize* the variance as much as possible with responsible aiming decisions. A good aim doesn't turn a poor loudspeaker choice into the right tool for the job, but we always want to be working to minimize the variance as much as practical within the given constraints. Even that amount of effort will consistently yield better results than the people who haven't bothered to take such an approach.

Another thing that can help is to gather some measurement data to quantify the level and tonal variance exhibited by the rig - not because we are able to change it, but so we know what it is.

The figure below shows four measurements taken on axis to one side of a "speaker on a stick" rig in a large theater from front to back - not nearly enough "rig for the gig" but that's what there is.

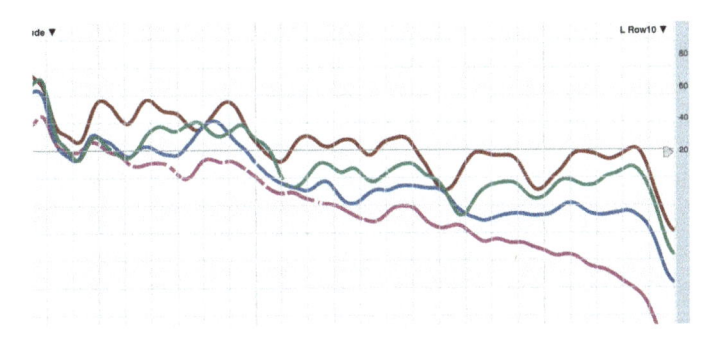

Figure 18 - Four measurements front to back on axis to a "speaker on a stick" in a large theater. (Smaart v8)

The measurements very clearly indicate significant level and tonal variance from front to back (guess where mix position is...). We can take this information under consideration during the operation of the event and be more informed about what's happening elsewhere, even if we can't do anything about it

Line Array Aiming Considerations

My starting point with setting array splays is to adjust the splays until

BETWEEN THE LINES /57

the centerlines are touching down at equally spaced intervals from back to front. For a typical audience geometry, this will result in smaller splays towards the top of the array and larger splays towards the bottom. Evenly spacing the centerlines helps to create an even distribution of high frequency energy from front to back and reduces the amount of shading we will have to do later.

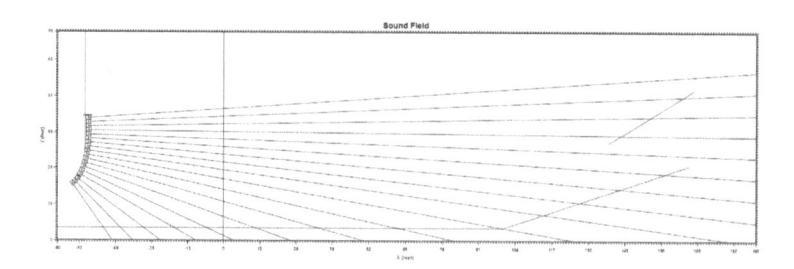

Figure 19 - Centerlines evenly spaced from front to back (MAPP XT)

I try to avoid any large jumps between splay angles even if it seems to make sense with the geometry. For example, I wouldn't jump between 1 degree and 5 without softening the transition with a 3-degree angle in between, at the very least. Keeping the angles as a gentle, gradual expansion rather than sharp jumps will reduce unpleasant interactions and lobing at high frequencies. (This high frequency disruption has been termed "shock" by Bob McCarthy.)

We must remember that a line array cabinet is not a laser gun. It's easy to visually trick ourselves with the center lines in a 2D prediction view into thinking we can "skip" balcony fronts and other architectural details, but it seldom works so simply in practice. In general, we're creating a smooth fan of energy, not an extremely narrow beam. Maybe that big 6° splay looks like it's skipping over the balcony front in 2D prediction software but of course in actuality there's energy between those two center lines.

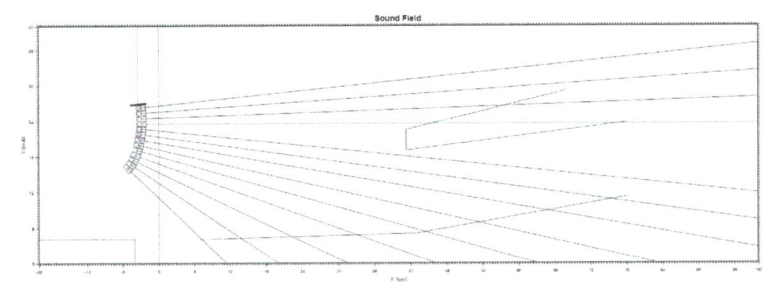

Figure 20 - We skipped the balcony front! (MAPP XT)

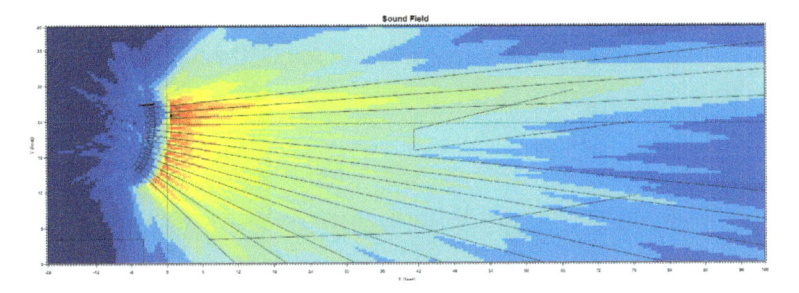

Figure 21 - Or did we? (4 kHz) (MAPP XT)

A 3D prediction will reveal the real behavior. Not exactly what we'd hoped for, is it?

Figure 22 - No balconies were skipped in the making of this prediction (4 kHz) (MAPP 3D)

Depending on the geometries, you may be able to slightly reduce the amount of high frequency energy hitting a wide balcony front with a "skip" in the splays, but this comes at the cost of increased coverage disruption and ripple above and below. Not all prediction software can render in high enough resolution to display it accurately.

The next thing to keep in mind is that the 2D side view (section view) prediction only shows us what's happening on axis at the center lines, and we must be mindful of our entire horizontal coverage shape, whether that be 90°, 100°, or whatever the elements are producing. We might try to "skip" a balcony front with the on-axis center lines, but further off towards the edge of the array, we've now created a cold spot in coverage that is landing in the audience area above or below the balcony front. If we are intentionally creating a design that puts a gap in our coverage, we need to be certain where that gap is going to land.

We have an additional complication here as well: unless we are aiming straight on into the front of the balcony, we have a dispersion pattern that's either aiming downward or upward at an angle. When an array projects energy onto a non-perpendicular surface, the coverage will either bend upward (if it's aiming upwards) or downwards (if it's aiming downwards) into a "smile" or "frown" shape. Thus, successfully skipping the balcony on axis to the array might also mean creating a gap in coverage into the audience members seated on the sides of the coverage pattern. These types of geometries can create some unexpected coverage patterns as a result, and a 3D prediction tool does a much better job showing these situations for what they really are.

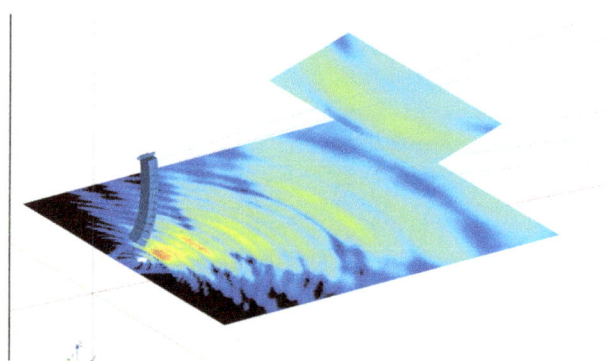

Figure 23 - Frowns on the floor, smiles in the balcony (MAPP 3D)

60 \ *Chapter 9: Design Considerations and Workflow*

One final design point is sufficient overshoot. Think about how a point source loudspeaker begins to roll off its high frequency output towards the edges of its horn pattern. The same behavior occurs with an array as well. If we aim the top box directly at the most distant seat we need to cover, that seat will end up outside the main high frequency coverage fan, but rather be "on the edge" and well into the high frequency rolloff.

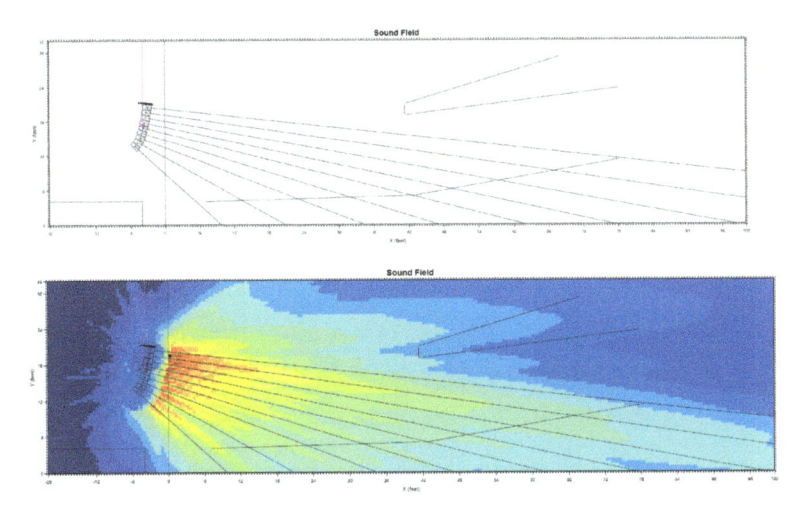

Figure 24 - With the top box aimed directly at the last row of seats, those listeners fall outside the high frequency coverage lobe (4 kHz) (MAPP XT).

If we want to adequately cover those distant listeners, we need to "overshoot" a few boxes of the array, which will help keep those seats within a more representative portion of the high frequency coverage. Two boxes of overshoot is a good rule of thumb to start. If you're up in the last row, and looking straight down the grille of the top box (or worse, looking down onto the rigging bumper), you're missing some seats at high frequency. Perhaps I'm starting to get repetitive, but prediction will indeed tell us what we need to know.

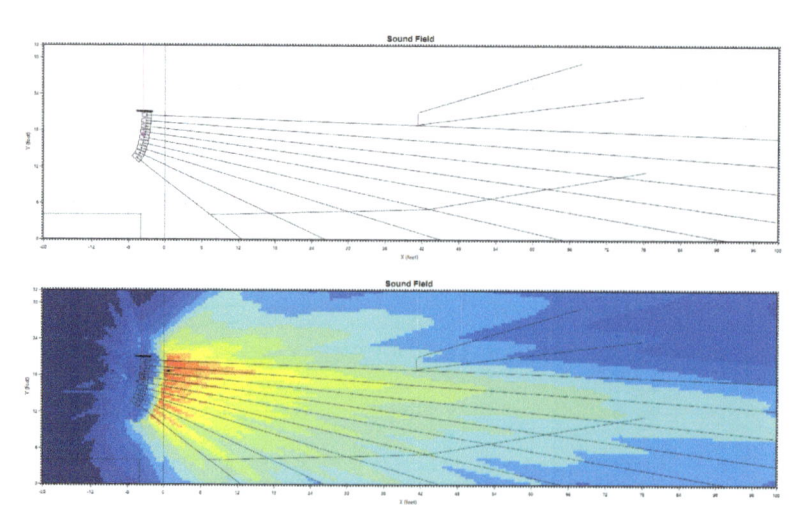

Figure 25 - With two boxes of overshoot, the last row is properly inside the high frequency coverage (4 kHz) (MAPP XT)

Trim height is a parameter that we often don't have a lot of latitude with, and so the expedient answer is "as high as possible," to help reduce the range ratio and get the bottom boxes of our array out of people's faces. But in a larger venue, an arena with high steel, or a large outdoor stage, etc, where we really can fly as high as we might reasonably want to, how high is too high? Is it possible to have too much of a good thing?

Lower trim gives us better perceptual imaging towards the talent on stage, whereas higher trim reduces range ratio. There's a happy medium to be had. There's an effect on the geometry as well: if our listening plane is a flat floor (or a flat outdoor field, etc), lower trim heights decrease the total vertical coverage angle that our array needs to cover, and can compress to such a degree that all our angles end up at 1°. If we splay open much more than that, we're oversplayed down front. The SPL is concentrated into a smaller coverage area, and the low-mid beam from the array is probably closer in direction and coverage angle to the coverage shape made by the high frequencies.

As we trim higher, we can open our splays up more due to the increase in total vertical coverage angle, which allows us to use a progression of splay angles more effectively as a mechanism to counter range ratio and decrease level variance. An increase in height usually brings with it an increase in uniformity. If you can't get enough boxes (not enough "rig for the gig") this might exacerbate things by increasing the required vertical coverage angle past what you can effectively cover with the given number of elements.

We also must consider the mechanical ramifications as well. As our array becomes increasingly curved, the center of gravity moves rearward. Once the center of gravity falls behind the rigging frame, we can't hang the array without a pullback rigging point. Even when the center of gravity lies beneath the rigging frame, a highly curved array can transfer most or all of the weight to the rear rigging point while leaving the front rigging point very lightly loaded, which may or may not be acceptable. (This is a decision for the qualified riggers, not for us.)

As trim height increases, the array needs to aim more steeply downward to hit the same coverage start and stop points, and the downtilt further shifts the center of gravity rearward. At lower trims, we will have more uptilt, which moves the center of gravity forward. When hanging an array requiring a lot of curvature, it can sometimes be difficult to achieve enough curvature to cover all the way down to the front of the audience area while keeping the center of gravity under the frame if we don't have a pullback rigging point. Decreasing trim height a bit means more uptilt to aim towards the furthest seats, particularly in a tiered seating geometry like a theater or arena.

That increase in uptilt can serve to shift the center of gravity far enough forward that we can get a few extra degrees of curvature at the bottom and still remain "riggable" and safe. Remember that prediction software can tell you everything you need to know about center of gravity, weight, and weight distribution, and as a general rule, we shouldn't attempt to hang anything without checking the weight and loading in the software to make sure it's safe. Most manufacturers' prediction tools will evaluate designs for weight limits and safety factors.

Yet another seldom considered factor is the rear radiation of the array. Trimming higher with more downtilt steers the rear low-mid lobe energy up over the stage, whereas a lower, flatter deployment directs more rearward energy towards the stage or the venue's back wall. Often, the energy leaving the sides or rear of a system is just as critical as what's coming out the front, so rear radiation might serve to inform a trim height decision in a sensitive application. We will revisit this concept of "360° thinking" during the tuning process. To summarize my approach for array splay, I am looking to satisfy three conditions as a starting point:

1. Two boxes of overshoot at the top of the coverage area
2. Center lines that touch down evenly spaced along the coverage area
3. Enough curvature to cover all the way down to the front of the coverage area

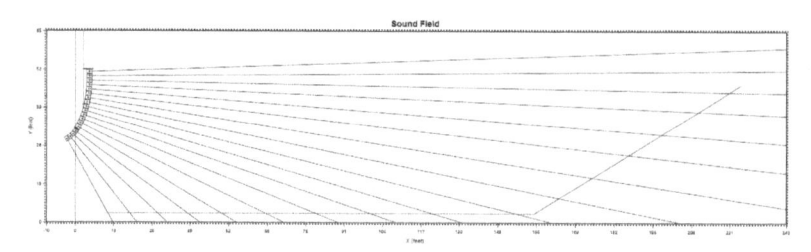

Figure 26 - Two boxes of overshoot, evenly impacting center lines and enough curvature to get down to the front. (MAPP XT)

I start from the top and work my way down, setting splays to achieve evenly spaced impacts as I go. If I get down to the bottom and have an oversplay condition, I can go back and tighten the angles up. If I run out of splay before I get down to the front of the coverage area, start again with wider splays.

Since we have small angles up top and large at the bottom, it's easy to figure out how much we need to adjust based on the amount of over- or undersplay. For example, if we end up with 6 extra degrees of coverage at the bottom of the array, changing the splays from 2-2-2-3-3-3 to 1-1-1-2-2-2 will contract the array by 6 degrees and now our bottom boxes should be right on target.

Subwoofers

A dose of reality is in order when dealing with subwoofer designs, as the best positions for optimal coverage are often not available to us. Flying subwoofers results in much greater uniformity front to back and also greatly reduces the high SPL "danger zone" in the front rows that present excess hearing health risk. But most often, particularly for smaller scale events, placing the subwoofers on the ground is our only option.

In the next chapter we'll take a survey of the various mechanisms we use to create cardioid subwoofer arrays. A cardioid array is a powerful ally in the fight to control low frequencies, as it drastically reduces low frequency energy that is washing over the stage, bouncing off the upstage wall, and arriving (very late) back into the listening area. I typically will choose to deploy some sort of cardioid sub array by default unless the specifics of the gig require that I do not. Keeping the stage cleaner overall can have a surprisingly large effect on the perception of a mix being "clean" or "tight" in the house. Some artists appreciate it as well, although some do not – again, rearward radiation is a design parameter, so it's something we want to think about at the design stage rather than "wait and see what we get" during the tuning of the system.

I once worked on a bluegrass event and an AC/DC tribute event both held on back-to-back days, same venue, same sound system. The first night, we deployed cardioid subwoofers and the second night we didn't. Minimizing stage wash was deemed critical for accomplishing a clean-sounding bluegrass mix, while it would have made the AC/DC tribute musicians feel unsupported, and they much preferred feeling some of the sub energy on stage during their performance. We remind ourselves of the principal tenet of system design: send the sound where it's wanted, and not where it's not. With subwoofer arrays, line length in the horizontal plane is a major player in creating the proper dispersion.

Fills and Delays

Fills get to pick up where the main system left off and have the remainder of the audience area assigned to their custody. The key is to position and aim the fills in such a way as to make the relative timing relationship along the seam be something that will hold over as large of a space as possible. We want to arrange a geometry that will let us move through the seam area while keeping the relative time offsets between the sources as stable as possible, which is what allows our timing decision to hold for more than a single seat.

By way of example, let's consider the worst-case scenario: an under-balcony fill located behind the listeners, on the rear wall of the venue. This creates a real mess in terms of timing because as we move away from the mains, we move towards the fill, and so the rate of change of time offset over the space is extremely high. Not only does this cause some undesirable side effects (spatial comb filtering) but it means we have to choose a single location at which we want those two sources to be in time, and come to terms with the fact that they will be out of time *everywhere else* (hence the comb filtering).

If we can instead hang the fills off the front edge of the balcony, so they're in front of the listeners aiming back at them, we can expect a much smoother and less tumultuous overlap zone, because moving backwards in the venue means moving away from both the mains and the fills, and the relative time offset between the sources changes much more slowly as we move around. It's the placement and aiming decisions here that enable us to make meaningful time alignment decisions later in the tuning process.

In general, this works best if we consider fills and delays to be lateral or radial extensions of the main system. Side hangs in an arena expand the coverage of the mains outwards in an arc, and front fills in a line across the front of a stage fill in the center gap with a lateral coverage extension.

In all cases, the fill's job is to restore the target level and tonality in its custody area to match that accomplished by the mains in the mains' custody area. SPL capability might slide a few dB either way - sometimes we want front fills a few dB louder than the mains, and sometimes the delay towers at a festival should be a few dB lower, but in general we want to begin with a consideration of uniformity, choose a product that can produce the appropriate SPL, and let the specifics of the use case push us a few dB either way if it's logical to do so.

Find the holes - then fill them.

CHAPTER 10
Friends in Low Places
(Cardioid Subwoofer Strategies)

In this chapter we'll learn the mechanics behind cardioid subwoofer arrays, which are a major tool in the quest for directivity at low frequencies. Although many manufacturers offer cardioid presets that handle the necessary processing to turn a specific physical arrangement of subwoofers into a cardioid array, the physics behind it are simple enough that we can "hand roll" a cardioid array in a few minutes.

There are three basic methods to obtain a cardioid subwoofer array from two or three elements. These configurations can be used in multiples, and can scale up to use larger quantities of loudspeakers. Once we learn the building blocks, we can apply them in our designs as necessary.

Let's start this examination with a pair of subwoofers side by side. Although we're not exactly omnidirectional, it's a far cry from the directivity that our designs exhibit higher up in the frequency range, so let's see what we can do to create a "loud side" to point towards the audience.

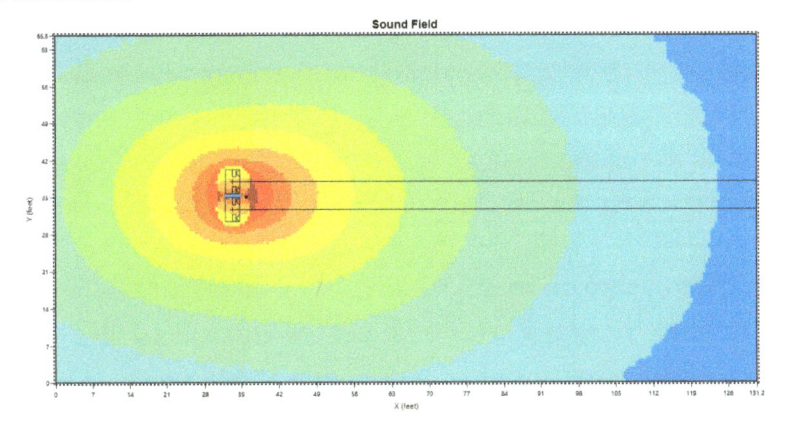

Figure 27 - Two subwoofers side by side, viewed from above at 63 Hz (MAPP XT)

Endfire

First let's look at the gentlest of the three methods: the "endfire" array (which identifies itself as "inline, non-inverted array" on its LinkedIn page to sound more professional). The endfire array works like a railgun: one element is placed closer to the audience than the other. The rear (upstage, furthest from the audience) box fires first, and the front box is delayed so that it fires as the energy from the rear box passes it. Thus the sources reinforce each other in the forward direction.

In the rear direction, we can build a cancellation that's about an octave wide at a frequency of our choosing by spacing the elements by about a quarter wavelength at that frequency. Let's aim for 63 Hz, which has a wavelength of about 17.85 feet, so we want an element spacing of about 4.5 feet. Remember that what we care about here is the spacing between acoustic centers – so we need to measure driver to driver, or grille to grille, or front edge to front edge, not the spacing between the cabinets.

Alignment is simple: we place the mic out in front, measure the rear box by itself, save the measurement then turn it off. Turn on the front box and adjust the delay on the front box until the phase traces line up. Since it's 4.5 feet closer, we'll probably need about 4 milliseconds of delay on the front box. (Common system engineer shorthand is to call the upstage, non-delayed box the "Time-zero" box, because it indicates the timing baseline for the rest of the array.)

Once that's done, we observe a summation at all frequencies in the forward-going direction. It's not going to be a perfect +6 dB, because one box is closer than the other, and so they're not exactly a summation of equals, but as we get further away from the array, the distance ratio approaches 1:1, so this improves with distance.

Consider what happens in the rearward direction. Let's go stand behind the array. From our perspective, the upstage (time 0) box is closer to us, and it fires first. Later, (4 ms later, to be exact), the downstage box fires, but then it takes another 4 ms for that energy to make it back to us, because it was both time delayed and farther away. So we don't see that energy until 8ms after the energy from the upstage box.

8 milliseconds is a half cycle at 63 Hz, and so at 63 Hz we have two sources arriving 180° out of phase with each other, building a cancellation that's about an octave wide... and that's about it. Since the cancellation is based on time offset (and therefore phase offset) it doesn't hold over frequency. In terms of rear cancellation, this is therefore the least effective of the three types of two-element cardioid approaches.

However, when used with 3 elements, delayed in a row, everything sums out front and arrives scrambled in the rear, building some significant rejection. If you've got multiples of three elements per side to work with, a 3-element endfire array can be a very effective solution.

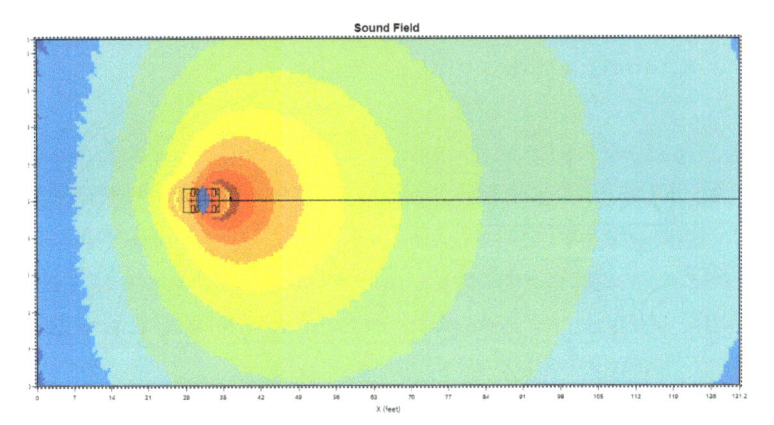

Figure 28 - Two-element endfire array at 63 Hz (MAPP XT)

Gradient In-Line

The gradient inline array uses the same physical layout as the 2-element endfire above: one element closer to the audience than the other, spaced by about the same amount as before. The difference is that this array is tuned from the rear, rather than the front.

Again, we imagine ourselves standing behind the array, however this time the front (downstage) box becomes the time-0 box, and we delay the upstage box so it fires as the rearward radiation from the downstage box passes it.

At this point we have the same configuration as we had before with the endfire, only firing in the opposite direction (louder behind). At this point we want to turn the rearward summation into a rearward cancellation, and we do that by inverting the polarity on one of the elements (traditionally the upstage, but either way you'll get the same effect). Now we have an array that is producing cancellation in the rearward direction.

What about out front? From the front, the upstage element is both late due to the delay, late due to the physical spacing, and out of polarity. So being a quarter cycle late by delay and a quarter cycle late by physical distance means we're about half a cycle late around 63 Hz out in front, but since we also have a polarity inversion in play,

BETWEEN THE LINES /71

we're now summing again.

We do start to lose some summation efficiency at lower frequencies because the delay and physical spacing mean less phase offset at longer wavelengths (which becomes a less helpful summation once the polarity inversion is introduced), but typically this is a tradeoff worth taking because we gain some extremely helpful rear rejection throughout the entire subwoofer range.

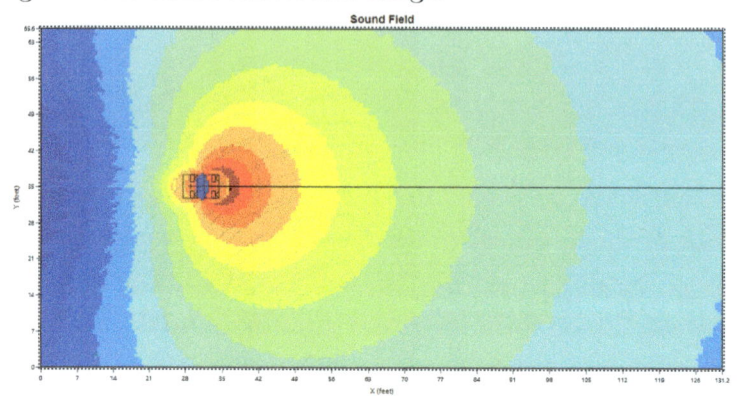

Figure 29 - Inline gradient array at 63 Hz (MAPP XT)

Gradient Stack

The gradient stack functions exactly like the gradient in-line array, except it has a smaller physical footprint because the two elements are physically stacked on top of each other, with one element reversed.

"But wait," you might object. "Haven't we spent the entire book talking about how subwoofers are effectively omnidirectional, and therefore there is no benefit to pointing one in the opposite direction?" This is an astute observation. I'm glad you are paying attention.

Reversing one of the elements does very little to affect the directivity on its own, but it does create some physical displacement between drivers. Or in other words, the acoustic centers of the two sources are no longer the same distance from the listeners out front. One is closer to the audience, and one is farther away, and it's this displacement that is critical to creating the directivity. These arrays rely on the physical displacement and the time delay reinforcing each other in one direction, and offsetting each other in the other direction.

A more common variant of this array uses three elements, with the middle element physically reversed, but you can get useful results with two as well. Same as the gradient in-line, we tune from the rear: measure the forward-facing (time 0 box), then turn it off and turn on the rear-facing box. Add delay until the phase traces overlay, then invert polarity on the rear-facing element. The time delay needed will probably be more than you might expect. That's because the energy from the forward-facing box needs to physically wrap around the sides of the cabinet before it can emerge at the rear, so it's later than the driver displacement alone might suggest.

For this same reason, prediction software can get you in the ballpark, but you won't know the exact delay time necessary unless you measure it in the field. Manufacturer cardioid presets have done this work for you.

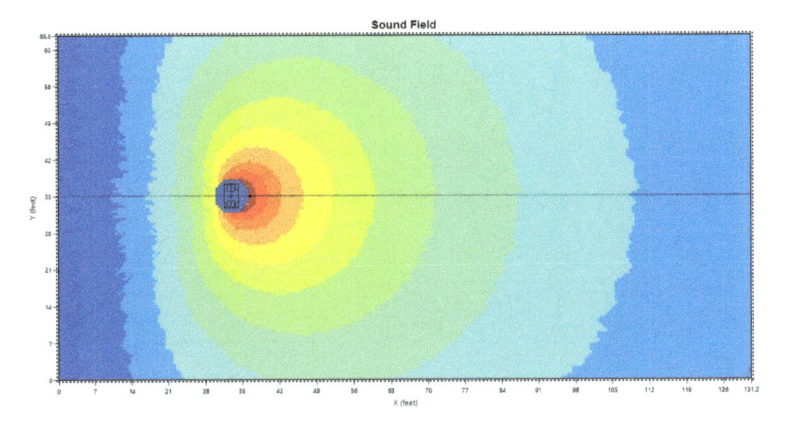

Figure 30 - Gradient Stack array at 63 Hz (MAPP XT)

Cardioid subwoofer arrays are one of the most powerful tools we have if our goal is to "point the loud side towards the audience."

SECTION 4
Perspective

CHAPTER 11
The Magnitude Fallacy:
Priorities, Tradeoffs and Free Lunches

Now you've sat through the opening act and are ready for the headliner to take the stage. The preceding chapters give us the technical context that we need for the rest of the book to make sense. This, however, is the "good part" - where we can dig into the conceptual basis for our decisions and use that information to get the best results for the largest number of listeners. This section is truly my "personal recipe," the techniques and methods that I have found most success with over the years.

I won't advocate for the supremacy of any approach explained within, beyond the statement that they have worked well for me. Take what you find helpful and leave the rest behind.

Number of Affected Listeners

English philosopher Jeremy Bentham (1748 - 1832) proposed as a "fundamental axiom" that "it is the greatest happiness of the greatest number of people that is the measure of right and wrong." Although we will steer clear of discussing the deeper philosophical implications, we will let the tenet of creating the most benefit for the most listeners guide us when we need to make a judgement call. (We will also steer clear of discussing Bentham's dying request to be dissected, preserved, clad completely in black, and seated in his favorite chair in a pose of deep thought, although the touring sound engineer is bound to encounter stagehands in a similar state.)

For example, consider a theater design with several underbalcony delay speakers. Each delay can be time-aligned to its adjacent neighbors, or to the mains, but we must pick one or the other. Some consideration at the design phase will minimize the disparity here, but often we don't get to choose, and we must use the positions that have been installed.

Time alignment is only truly meaningful to listeners who are in the coverage of more than one source (including any acoustic energy from the performance on stage). We can therefore figure out where we should focus our time alignment efforts by identifying audience regions where the listeners are in the coverage of multiple sources.

How many people are sitting in the overlap between two adjacent delay fills? How many people are sitting in the overlap between the mains and a delay fill? We want to prioritize whichever seam has more listeners in it. If there are 150 seats that are covered by both mains and delays, and only 15 seats on the seam between two delays, it's clear that the main-delay timing should be the priority since far more listeners will experience the consequences of our timing decision there, and the delay-delay timing will land where it lands. In practice, the delay-delay time offsets will probably be very small once each is properly timed to the mains, and so we can likely get the best of both worlds, or something close to it.

If we have an entire delay ring that shares a single DSP output, and we must decide on a single delay time for the entire ring, we'll want to delay it back far enough that *none* of the delay speakers arrive at their custody areas before the mains. We want the listeners seated in the overlap to localize to the stage, so delays should either arrive on time or late, never early. (A delay arrives precisely when it means to.)

Two milliseconds of extra delay at the seam is sufficient to shift the perceptual localization to the earlier source, but not enough to cause flamming or smearing that can get us into trouble with larger time offsets, particularly at high frequencies. Once time offset reaches about 10 ms or more we're crossing into the region where it can become quite problematic. If our delay ring has a 10+ ms span of alignment times that would be needed to align it back to the mains, and there's no way to break it out into more processing channels, we might need to split the difference so everything arrives as close as possible to the arrival from the mains - "center the error bars" you might say.

In that case, odd localization from a slightly early arrival is better than suffering a loss of intelligibility because *it's so late*. The logic driving this decision is the same: within the constraints of the design, we want to make decisions that make as much improvement as possible for as many of the affected listeners as possible. Conversely... choose the option that has the least amount of fallout for the smallest number of listeners.

So when we need to choose between two options, one factor we always want to consider is the number of people affected by each solution, and which option would lead to a better result for the most listeners.

Main-Sub Alignment

It is a perennial source of some amusement (and some bemusement) for me to see how much mental bandwidth the community at large dedicates to Main-Sub alignment. This is certainly the most "famous" of alignments, and the only one that students predictably ask about, perhaps because it's the only one that remains in the hands of the end user.

With modern systems, manufacturers have taken care of the woofer-tweeter alignments and so forth. Anything that resides within the same loudspeaker cabinet is safely out of the hands of the end user and sorted out for us by the manufacturer's design and processing presets. Perhaps it arises from the need to feel like we've "done something" to the system.

The elephant in the room with main-sub alignment is that in many cases - one might even say, the majority of cases - it's ignored during the design stage to the point that it doesn't make much sense to be much concerned with it during the alignment stage.

If the subs are on the ground and the mains are flown, those two sources are displaced from each other such that there is no one correct alignment time for all listeners. As you move throughout the space, the relative arrival times from the subs and mains will shift enough that we need to pick a location at which we want the subs and mains to be phase-aligned (only throughout the frequency range where they both contribute energy at similar levels, mind you) and let them fall out of phase elsewhere.

If the main-sub alignment is important to you, you should design your system in a way that allows the alignment to be a meaningful decision that is preserved over the listening area (fly your subwoofers adjacent to your mains).

Let's consider an example. A typical large club situation, mains flown at 20 feet, subs on the deck, front row of listeners 10 feet from the stage and furthest listeners 90 feet from the stage. Up front, the subs arrive much earlier than the mains (path length difference of approximately 12 feet). In the rear, the mains and subs arrive much closer together in time (path length difference approximately 2 feet).

The time offset varies over a range of about 10 ms depending on where we choose to stand. At a crossover frequency of 100 Hz, that means the main-sub relationship is experiencing *every possible phase offset* somewhere in the venue. As I like to say, if you don't like the main-sub alignment, choose a different seat.

Now, keep in mind - we only concern ourselves with energy in the frequency range where the mains and subs overlap. On a small or medium format system this might be a pretty small frequency range indeed - perhaps less than an octave - and moreover, the relative levels are going to change depending on proximity. As we move forward in the space, the subs will get louder *faster* than the flown mains, because we get closer to them faster. The frequency range of overlap might be sufficiently different in the front of the room compared to the rear that we end up with a completely different crossover interaction (both in frequency and time). Want to fix this? Fly your subs next to your mains. Now they're coupled and will maintain their relative timing relationship much more tightly as we move around the space.

We must also be mindful that frequency-domain data (magnitude and phase) doesn't show us the whole story. Time-domain data reveals a more complex reality. Take an impulse response measurement of the system and bandpass the resulting data to the sub crossover range, say the octave band centered on 80 Hz.

You are likely to see a rather lumpy, undefined mush of energy arrival rather than the clean peak such as we are used to seeing at higher frequencies. Subwoofers have non-trivial amounts of time smearing due to the mathematical properties of acoustically and electronically bandlimited systems, and the floors, walls, and ceilings of our venues do their part by adding a cornucopia of late arrivals.

All of this taken together means it's not uncommon to see subwoofer energy arriving at the measurement mic over a range of 80 milliseconds or more, even in a relatively small venue. Since our subwoofers lack directivity, energy goes *everywhere* and a lot of it eventually finds its way to our measurement mic position after bouncing off of room boundaries. This can make the concept of "arrival time" a very slippery notion indeed.

The phase trace (frequency domain data) doesn't really give us an adequate appreciation for the time-domain mess that is typically occurring in that frequency range inside a venue. Measure it and see for yourself, then decide how much of your time you think it's worth .

Main-sub alignment gets the blame for a lot of the perceived lack of "tightness" in the sub range but in reality, the majority shareholder there is often the fact that energy in the sub range is arriving over such a long period of time rather than all at once. In my view, the best thing that can be done (aside from acoustic treatment, which isn't an option if we are guests in the venue...) is to deploy some manner of cardioid sub array, which helps keep more of the subwoofer energy directed forward into the listening area, and less hitting the walls and arriving later into the listening area.

The upstage wall behind the band is a main culprit, and using a cardioid deployment to reduce its contribution by 15 dB or more can make a significant difference in reducing the magnitude of that late-arriving energy, and also in the corresponding perceived "cleanliness" or "tightness" of the system at low frequency.

I will almost always deploy cardioid subwoofers unless I have a reason not to, and I will spend enough time investigating the main-sub interaction to make sure that the FOH mix position isn't suffering some sort of cancellation, but beyond that, I will only give it as much mental bandwidth as the design allows it to occupy in any meaningful way.

Aux-Fed Subs

Let's expand on the previous discussion and consider the practice of driving the system subwoofers off a separate console feed, whether that be an auxiliary bus, a group, or a matrix.

The common objection to this practice is that by varying the subwoofer send level, one is effectively shifting the crossover frequency between the mains and subs, and this causes angry witches to be released from the amplifiers and cast spells upon the listeners (or something similar, I might not be remembering correctly).

As we've seen from the discussion above, the idea of subs-on-the-ground and mains-in-the-air having a uniform frequency or time relationship is often rather unfounded to begin with, so how much concern does this warrant?

It's very easy to quantify this issue for yourself in a given situation. Once you've measured your main and subwoofer responses, add a trace offset to the subwoofer trace in the analyzer - slide it up and down on the screen by 6 dB either way and see how much the crossover point changes (and remember that response is probably different somewhere else in the room). In Smaart®, you can offset a trace by using the dB + and dB - buttons on the command bar, pressing [Ctrl/Cmd] + [Up/Down] arrows, or simply clicking and dragging the trace up and down.

It depends on the specifics of the loudspeakers in use and the slope of the crossover filters, but typically a 6 dB level change in either direction might shift a crossover frequency by a dozen Hertz or so. If you were aligned before, you are probably still aligned.

This is a wonderful example of what I have christened the "Magnitude Fallacy" - something that is conceptually true, and has a factual basis in real science, but the scale of the effect is much smaller than people might think. "Yes, it's true, but no, you probably shouldn't worry about it."

I once wrote a ProSoundWeb article whose title, "Don't Phase Me, Bro" was one of the crowning achievements of my career. The article explores this idea in more detail, and suggests that the concerns of "phasing issues" in these situations might be unfounded. (The phrase "phasing issues" is typically a red flag that some manner of unsubstantiated statement will soon follow.)

In the follow-up article "Don't Fear the Phase Gremlins" (are you getting the idea yet?) I described a test scenario conducted as part of a Smaart® training in Brooklyn NY.

The training took place in a small theater that had good acoustic treatment, and main arrays with the subs flow directly above the mains, so we had a situation where the main-sub alignment was cohesive enough to spend some time getting it right.

We first worked to achieve the optimal main-sub alignment (which, to no one's surprise, ended up being the manufacturer's default presets). I then made the main-sub alignment as *poor as possible* by inverting the polarity of the subwoofer and turning a perfect summation into a cancellation. The result was a -5 dB dip about 20 Hz wide. Seat-to-seat variations in the sub range tend to equal or exceed this level of tumult, so even when it's *really wrong* it's not likely to be a show-ruiner.

Two additional lesser-discussed points about the aux-fed-sub controversy. First – *not* sending an input to the subwoofers is the most effective high pass filter one could ask for – except when it's not. Over the years I have noticed a significant correlation between mix engineers who prefer a significantly tilted system response and those who request aux-fed subwoofers. When the system has a severe LF tilt-up or "haystack," it can "undo" the effects of the input channel high pass filter (Yes, you rolled off at 18 dB per octave, but you have an 18 dB sub haystack, and they're fighting each other).

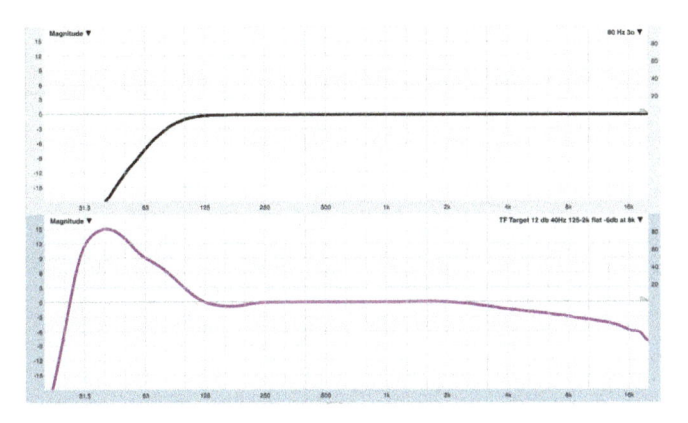

Figure 31 TOP - Console High Pass Filter. BOTTOM - Example system target curve (Smaart® v8)

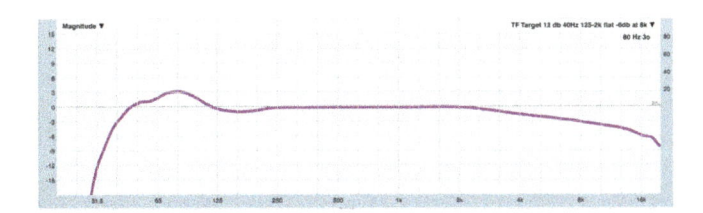

Figure 32 - The combined response (HPF + System Tonality) (Smaart® v8)

Mix engineers with less tilted target responses tend to see their console high pass filters being more effective as a result, and are less likely to request an aux fed sub configuration for their systems. It's a mixing tool whose utility is directly related to the amount of tilt in the target curve.

BETWEEN THE LINES / 85

I don't have a horse in the race either way – the mix engineer will be given whichever they request – but hopefully it is an informed decision that is based on understanding the tradeoffs, however severe they may or may not be in a given situation.

On a large-format system, the mains often have low-frequency extension well into the sub range, down to 50 Hz or even lower. In these cases, the subwoofers are adding additional low frequency headroom and extension but are not the sole custodians of the low frequencies. In a flown configuration such as typically encountered in arenas, stadiums, and festivals, removing an input from the subwoofer send doesn't prevent the system from reproducing that input's low-frequency content, because the mains are still in the game down there.

Instead, the directivity of the low frequency energy from that input could change, because many designs rely on the time / phase relationship between flown mains and flown subs to create directivity. (A flown main hang with a flown sub array directly behind constitutes a 2-element endfire array in the horizontal plane.)

Typically, aux-fed subs are requested by mixers who have been working in smaller clubs, where a Subwoofer fader is a very quick and functional way to quickly control the overall tonality of the system, although there are better ways in my opinion, such as a simple EQ on the Main LR matrix driving the PA.

When their act moves up to bigger venues with large-format systems, the aux-fed sub will not behave the same way they are used to (a system tilt control on a fader), and with systems that large, there is probably a dedicated systems engineer to address tonality issues for them, so that control is less likely to be needed. The systems engineer is (hopefully) providing a system that has consistent tonality from day to day, so once the board mix is balanced as desired, it should carry through.

It's all about priorities – make sure you give things as much of your attention as they deserve – but no more.

CHAPTER 12
Much Ado About Microphones

Microphone Correction Files

Now let's tackle some measurement microphone mythology. (Apologies to alliteration enthusiasts, that's as much as you're going to get). Microphone correction files are a common source of anxiety amongst people who are early on in their measurement careers.

The concept is that microphone manufacturers will individually measure each measurement microphone they produce in a controlled environment, and issue a data file describing the microphone's deviations from a flat response. This file can then be loaded into the audio analyzer to correct for those deviations.

Some lower-cost measurement microphones either do not include such a file, or include a generic file that describes common deviations for that model of microphone. Concern is often voiced that such a microphone is then unsuitable for measurement work because the file is missing or inaccurate.

In practice, even inexpensive measurement microphones are nominally flat from 40 Hz to 4 kHz. Extending that flat response down and up into the last few octaves of the audible spectrum requires better quality mics with closer manufacturer tolerances, but this is still a feat that can be achieved by even moderately priced microphones.

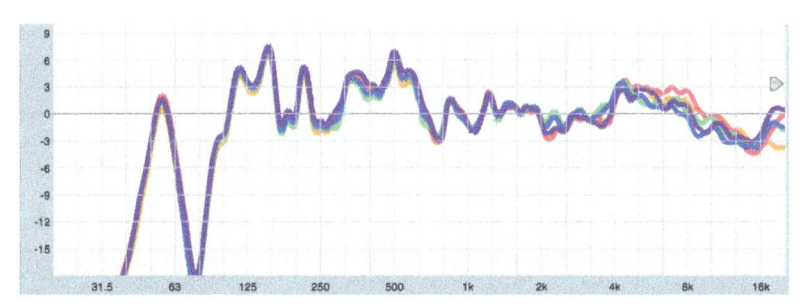

Figure 33 - Five microphones made by four different manufacturers, measuring the same source, with no correction files (Smaart® v8)

The figure shows five measurement microphones made by four different manufacturers spanning a price range from $80 to $750, a 10:1 price ratio (or 20 dB if you did your decibel conversion flashcards). Preamp gains have been adjusted to account for the differences in sensitivity, and as we can see, they all give us virtually the same answer. The only deviations worth discussing exist in the top octave and are still small enough that I would feel comfortable using these microphones interchangeably in the field.

This "mic compare" routine is a good habit to get into before you start a multi-mic measurement session. Put all the mics in the same place, turn the system on, adjust the preamps until all traces overlay. If a microphone is broken, you'll see it.

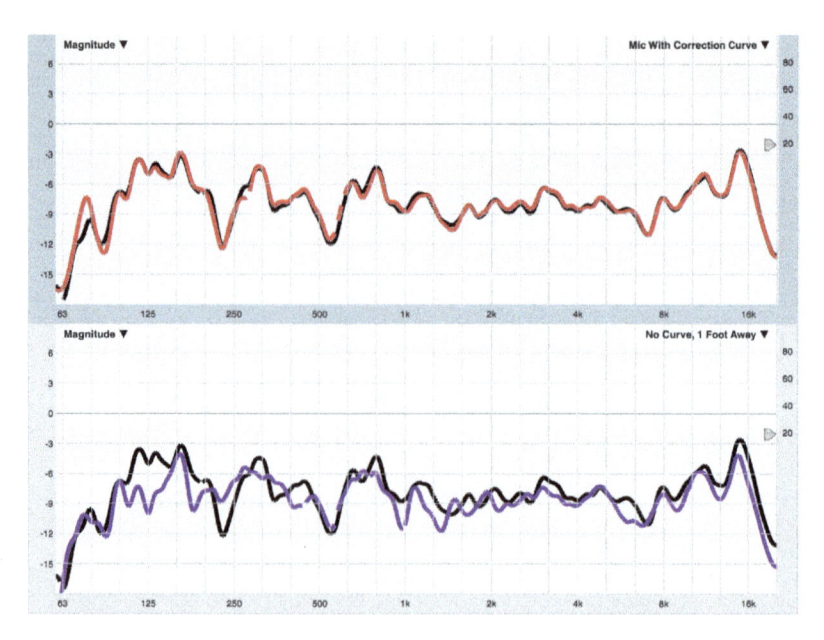

Figure 34 TOP - Two measurements taken from the same location, with and without the mic correction file.
BOTTOM: Two measurements separated by a foot. (Smaart® v8)

The top of the figure shows two measurements taken of the same loudspeaker at the same mic position: with and without the microphone correction file loaded. The bottom pane shows the same mic position vs another measurement taken a foot away. The seat-to-seat variations in the venue absolutely swamp the small deviations that are addressed by the correction file.

Our field measurement work doesn't happen in laboratory environments, and microphone correction files are another example of the Magnitude Fallacy at work. This is a real thing, but it's not something you need to worry about. If your microphone is of high enough quality to come with an individual correction file, it's probably of high enough quality not to need it. However, there is a less obvious benefit to purchasing a mic that includes a correction file: reassurance that the manufacturer has tested the microphone and confirmed that it is performing within its specifications.

Matched Microphones

Another related point of concern is whether one needs to purchase "matched" measurement microphones. Hopefully the above discussion has convinced you that even different models from different manufacturers should match within a reasonable tolerance, so you might reasonably wonder what benefit a matched pair or matched quad brings to the table.

From a measurement microphone manufacturing standpoint, "matched" typically refers to the microphone's sensitivity. For a given SPL at the diaphragm, what output voltage is produced? That's sensitivity, and typically this can vary a dB or two either way from mic to mic even of the same model.

This doesn't concern me because I have a preamp I can adjust until all mics produce traces at matching levels. If you purchase a matched pair, you can be reasonably certain that both mics will produce overlaying traces when the preamps are set the same. (But you did confirm that for yourself, right?)

For measurement purposes, I'm not convinced that the few seconds saved adjusting a preamp is worth the additional cost for a matched pair, however if you plan to use those microphones for other tasks such as recording, that may be an investment that is worthwhile to you.

Are All Mics Created Equal?

You might feel that, given these recent and totally scandalous revelations, there's no reason to buy expensive measurement microphones. However, there are substantial differences between models in areas like build quality, longevity, and unit-to-unit variance. If you're spending under $100 or so on a mic, it's probably not going to live a very long life - some of those cheaper models are held together with nothing but hot glue. This is probably just fine if it's your first measurement mic, or if you are working in hostile conditions such as a dusty outdoor festival for the simple reason that if it breaks, or gets knocked over, or smashed on the arena floor, or run over by a forklift, it's a minor inconvenience and not a major financial loss.

$250 - $350 USD is a good budget for investment in a professional, reliable measurement microphone. This way, you know you have at least one good-quality, well-built microphone that you can trust and can be used as a performance check against all your cheaper mics. A good measurement mic will serve you well for years to come.

Diminishing returns is a real factor here, and for our purposes, it might be a better investment to buy two $350 microphones than a single $800 microphone.

Another technical point: measurement mics have varying sensitivities and Max SPL ratings. These factors don't really come into play for typical measurement purposes but for those seeking to measure SPL data at concert levels, some due diligence is required. Make sure the mic can tolerate the high sound levels (Peak values in the mid 130's) and that the sensitivity isn't too high as to overload the connected preamp at those levels. At such high sound pressure levels, all measurement microphones are not created equal, so if you're planning to do concert-level SPL measurement, it's best to check with the audio analyzer manufacturer's support team to ensure you're using the recommended equipment.

Spend enough on your measurement mic so that you don't have to worry whether it's telling you the truth.

CHAPTER 13
Splay and Pray

We now turn our judgmental gaze away from measurement microphones and towards two specific array splay-related topics that deserve some examination.

0° Splay Angles

The first is the practice of setting multiple cabinets in a line array to 0° splay. This is a practice that is thankfully becoming less common, since some manufacturers have caught on to the fact that it doesn't work as well as everyone seems to think and have started releasing cabinets that don't allow for a 0° splay setting.

Splay angles between cabinets control the vertical dispersion of high frequencies in an array. We design our array splays to create a fan-shaped coverage pattern that widens over distance and serves to cover our audience geometry. When multiple elements are splayed at 0°, they are now 100% overlapped, and this overlap mechanism works to create a coverage shape that *narrows* over distance at high frequencies. "The array that splays together stays together."

This behavior has been described in works by Harry Olson and others, and is probably been best characterized by Bob McCarthy as the "parallel pyramid" mechanism. A high frequency coverage shape that narrows to a point over a great distance is problematic enough, in that it doesn't match the shape created by the curved portion of the array, which widens over distance, creating a coverage gap.

Worse, the tip of that high frequency spike will land at a different distance from the array over frequency. This creates what I consider to be a non-optimizable coverage shape that can't be fixed with EQ except at a single location (because EQ can't change coverage shape), and we've given up the fight for uniformity right out of the gate, with an array geometry that changes coverage shape over frequency.

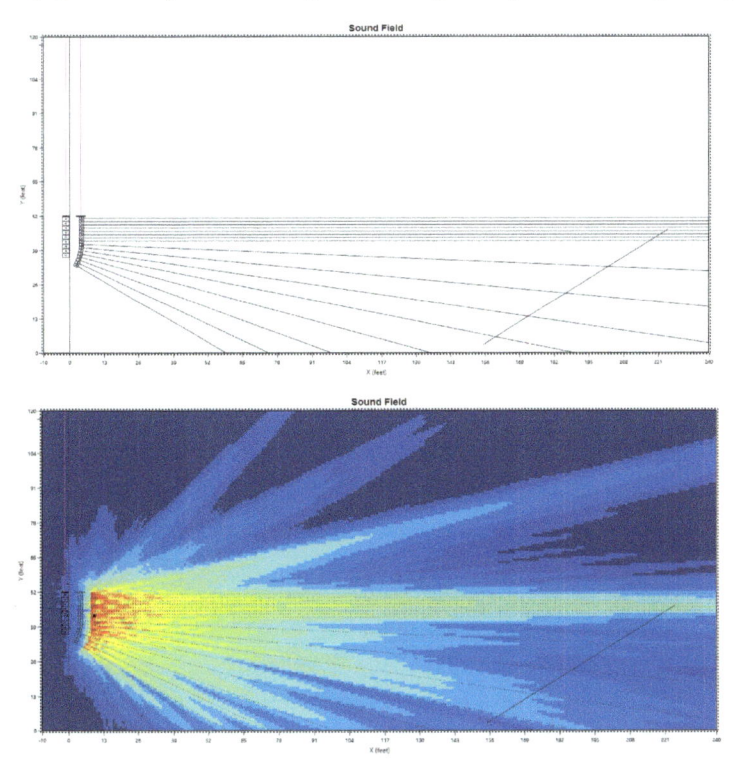

Figure 35 - The narrowing dispersion and high frequency gap created by 0° splays (4 kHz)(MAPP XT)

The common justification for splaying a bunch of boxes at 0° is the obvious one - "we need it to throw further." If you've ever been far out at a festival or in the back row at an arena show, and heard a harsh, "bandpassed" sound that feels like a loud cellphone speaker, you were likely listening to a number of elements splayed at 0 degrees.

The physical dimensions of the array cabinets prevent large numbers of elements from summing constructively at the highest frequencies, further exacerbated by air absorption. The lower frequencies are falling off at this distance because of the inverse square law behavior, so we are left with an unpleasant, nasally bandpass sound in the coverage of these boxes that typically cannot be satisfactorily addressed with EQ. Of course, EQ can change the tonality of a signal but can't change the fact that the coverage shape narrows over distance, and it's that narrowing that is problematic here. (Remember the bit about making the same coverage shape across all frequencies?)

A flown array will usually put compression on the front rigging pins and tension on the rear rigging pins unless a pullback or compression rigging arrangement is used. There is some amount of "slop" in the rigging (otherwise you wouldn't be able to get the pins out) and when the array is flown, the angles can slightly compress. If you were already at 0°, congratulations, now you're concave, which sounds as bad as it sounds. Using a pullback rigging point, or compression rigging, can alleviate this issue.

My policy is "just say no." If I can't achieve the necessary throw without resorting to 0° splays, I either need a more powerful loudspeaker, or delays (or both). Sub-1° splays such as 0.5° or even 0.25° (which can essentially become a 0° splay once hung) can sometimes still result in undesirable beam narrowing and often don't sound as natural compared to the larger splay angles. For this reason, I usually attempt to avoid sub-1° splays in my designs.

As a last resort, you can accomplish splay angles smaller than the manufacturer's smallest non-zero angle by staggering them. For example, if your options were 0° and 1°, you could try alternating between the two. 0°-1°-0°-1°-0°-1° would give you an effective 3° coverage angle with more SPL capability than hanging 3 boxes at 3°. Although this may still not be optimal in terms of sound quality, it will at least help to prevent the array from becoming concave when flown.

"That's not an array. THIS is an array!"

Another "greatest hits of internet audio forums" is the debate of how many elements must be in an array before we can properly refer to it as, or consider it to be, a line array, as if it's listening to us and might become offended.

As we have learned, decreased line length results in loss of directivity at lower frequencies, and in that regard, there may not be much benefit over a single point-source cabinet at 100 Hz.

However, even three elements can yield a significant benefit compared to a single point source cabinet in terms of high frequency directivity, and if our goal is to control variance, having two splay angles to adjust is defensibly better than zero in a lot of applications.

An example is in order - a graduation event held in a school's football stadium. The seating area geometry is a single constant rake. The vendor has placed loudspeakers on stands along the length of the seating area.

The point source box on a stand causes two issues: First, the range ratio in this situation exceeds 6 dB (farthest listeners are more than twice as far away as the nearest listeners), so we can't adequately counteract it using a single point source box. It will be louder in the front. Since this is a spoken word application, we want to go for as much uniformity as possible within the speech band.

Secondly, the speakers would need to tilt back to aim at the farthest row, and they can't due to the way they're mounted on the poles (cups, not yokes). We have a situation here where the listeners in the front row are directly on axis, and also closest, and as we move deeper into the listening area, we move both off-axis of the horn and farther from the loudspeaker, so we get the double-whammy of level loss. Measurements taken on-axis from front to back show us that the resulting level variance (12 dB) leaves something to be desired.

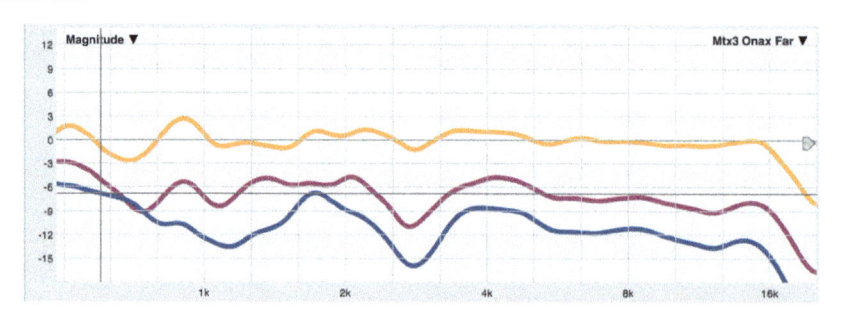

Figure 36 - Front to back variance in the speech range from the speakers on sticks (Smaart® v8)

The mix engineer has a bit of a tightrope to walk here, as "loud enough to be clearly audible" in the rear could be uncomfortably loud down front.

The next year, I was asked to help with the design. The company had purchased enough elements of a small-format line array product that I was able to use pole-mounted stacks of three elements at each speaker position.

Since the point source elements from last year were on pole cups, not yokes, switching to the line array elements allows the immediate benefit of being able to adjust the tilt upwards until we're aiming at the last row. We also have gained the ability to use splay angles and create whatever vertical coverage angle we desire at high frequencies.

Both benefits would also exist for a simple two-box stack, but a third box means two splay angles to adjust and that allows us to introduce asymmetry into the design to further offset the level variance, at least in the high frequency range – and this is a speech gig, so that's an improvement worth making.

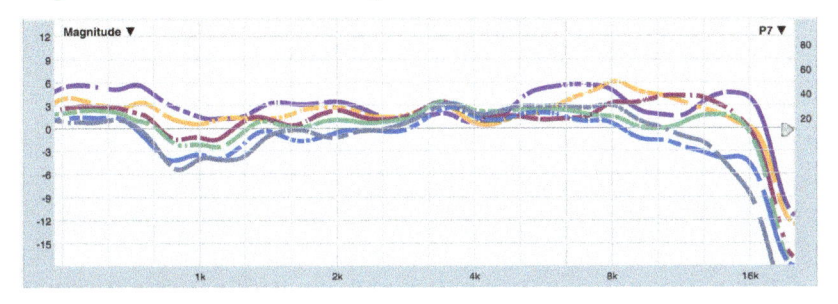

Figure 37 - Front to back variance in the speech range from the 3-element stacks. This plot uses the same vertical scale as the previous plot. (Smaart® v8)

And improvements have certainly been made. If we look at the measurements gathered in the same front-to-back spread, we see the speech band has taken a giant leap forward in terms of level uniformity, even before any EQ or processing has been applied to the system – this is simply the array mechanics doing their thing. The variance above 8 kHz is largely attributable to air absorption.

The mix engineer is free to adjust the overall tonality as desired, knowing that it will translate much better and they can have more confidence setting the SPL for the event without worrying about the level variance causing problems for listeners down front or up in the back.

I am far less interested in joining in the debate about what to call this type of arrangement, and far more interested in whether it can help me achieve my design goals and make meaningful improvements.

"Well you see, ma'am. The reason you couldn't hear is that someone on the internet said I can't use just three boxes."

CHAPTER 14
The Confusion About Tilt

The idea of overall system tonality - what we colloquially refer to as "tilt" - is a topic of ongoing discussion between mix engineers and system engineers. As we learned earlier, it is natural for systems comprised of multiple loudspeakers playing together to exhibit some amount of low-frequency emphasis or "tilt up" in the tonality.

Most professional mix engineers have a preferred "target curve," or tonality they would like the system to achieve so that their board mix will sound as they intend. This can vary tremendously across genre, program material, and individual mix engineers. Musical theater and corporate systems are often tuned quite flat, while rock and roll systems might have a 6 to 12 dB tilt plus an additional 6 dB of subwoofer "haystack" at the bottom, and EDM systems might be even more severely tilted.

The mix engineer will do whatever they need to in order to achieve the desired tonality at mix position. If the system sounds bright, they will cut back some high end. If the system sounds muddy, they will cut back some low mid. The question is where that filtering should be done - should it happen in the console, or in the PA tuning?

From a "signal transmission" point of view, we don't want the mix engineer to have to re-mix their show every night when they plug into a different sound system - we want to adjust the tonality of the system to ensure the board mix sounds as intended, not the other way around.

This doesn't mean that the PA system shouldn't make a tonal contribution to the mix - just that the tonal contribution should be consistent from system to system. In other words - you wouldn't EQ your reference music to make the PA sound right, you would EQ the PA to make your reference music sound right.

The target tonality of a sound system is largely a matter of preference, however there are some objective reasons to choose more or less system tilt.

When the PA system is contributing to the tonality of the mix, that means that when the mix is sent out of the mixing console to any other destination (an internet stream, for example), it will be missing the low frequency contribution of the PA and will sound unpleasantly thin.

Based on analysis of SPL logs from live music events, we can state with certainty that live music events typically have more low-frequency energy than recorded material of the same genre. Sound level measurements are commonly made with the standardized C-Weighting, which includes most of the full audible range, and A-Weighting, which de-emphasizes the low frequency content. We can examine the difference between the long-term C-Weighted and A-Weighted SPL values ("C minus A") to give ourselves a rough indicator of the amount of low frequency content in a mix. Depending on the genre, we can expect the live show to have 6 - 10 dB higher C-A than the recorded material, so it stands to reason that if our PA system supplies this amount of low frequency tilt, then the resulting board recording will have a tonal balance appropriate for recorded material when listened to in isolation, without the tonal contributions of the PA.

C-A doesn't directly correlate with system response tilt but as a rough guideline, perhaps 6 - 8 dB of tilt below 250 Hz and 6 dB additional "sub haystack" will land you in the ballpark with typical program material. A moderate amount of tilt yields a board feed that sounds tonally desirable without any EQ applied after the fact, which is a valuable commodity in many applications.

Another factor is convenience. System tilt can be thought of as "convenience EQ." If the system has a lot of low frequency tilt, it's common to see mix engineers using a lot of aggressive high pass filters and low-shelf cuts on their inputs.

It's easy to see why: think of capturing that tilted PA response as an EQ preset. Then flatten out the PA response completely, and load that EQ preset onto *every input channel in the console*. The result would have the same overall tonality, but this seems crazy, doesn't it? You'd want to cut a lot of that low frequency boost out. You can either do it once in the PA, or 32 times in the input EQs.

By that way of thinking, the optimal system tilt is the one that requires the mix engineer to do the least amount of EQ in the console. If we are systems engineers, the answer is simple: whatever the FOH mixer wants.

A less-often considered factor is headroom. If a large format PA exhibits a large amount of low frequency tilt by default, and we use EQ to remove some of it, we are effectively decreasing the voltage gain of the PA at low frequencies. This means we must drive the console outputs harder to get the system to the same SPL, and if the PA was underpowered to begin with, cutting back the low end by 14 dB might mean your console can no longer drive the system to the desired SPL without overloading the console outputs.

Uniformity is not Tonality

Hopefully it's self-evident by this point, but it bears a mention that uniformity and tonality are two different things. The tonality (target curve) of a system can be whatever the mix engineer wants it to be – our job becomes achieving that tonality across the entire listening area.

I have heard mix engineers say things like "I don't use Smaart®, because I don't like to mix into flat systems." The analyzer doesn't decide your system tonality (or anything at all, for that matter). It simply provides information. Since variance is something we can measure objectively, the analyzer can help us understand how consistently our system is achieving the desired tonality in various parts of the listening area. That information helps us decide what parts of the system we might want to adjust to increase uniformity.

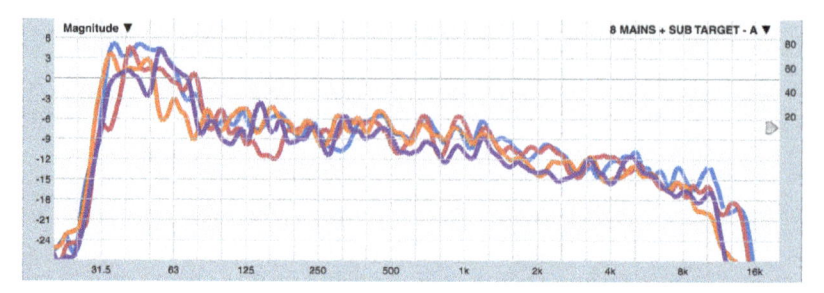

The figure shows measurements taken throughout the coverage of a system deployed in a 14,000 seat arena. The data indicates a very high level of uniformity, and a very non-flat tonality. (Smaart® v8)

If the mix engineer turns on their console, listens to material through the system and doesn't like the overall tonality, it is very easy to adjust - we can adjust the EQ until they are happy without any effect on system uniformity. If we've done our work with the design and tuning up until that point, we can be confident that the system will have the desired tonality everywhere else, as well.

> "I'll make it sound good right here. You make it sound good everywhere else."
> – Wayne Pauley

SECTION 5
Alignment

CHAPTER 15
Processing and Drive

The last thing we need to think about before we can tune our system is how to drive and process it. Hopefully we've done as much as we can with our mechanical choices (placement, aim, splay) to reduce variance, and create an arrangement of loudspeakers that will be straightforward and efficient to tune.

Processing will push us across the finish line. Besides any manufacturer-recommended presets for the particular loudspeakers we're using, we want to be able to control the level, timing and tonality of the various system elements and adjust them as needed in our quest for uniformity.

Although there are proprietary, manufacturer-specific tuning tools and processing algorithms on the market, we can accomplish the lion's share of this work with the standard "stock" processing options that are available in virtually all processors: gain, polarity, delay, and EQ filters.

As you gain familiarity with different platforms, you will develop preferences for feature sets or workflows, but keep in mind that no manufacturer has their own physics (if they claim to, they're lying) and that we can do highly effective work towards highly uniform systems with these standard tools.

Processing Granularity

In addition to the physical and mechanical limitations of the design, our ability to achieve uniformity is also limited by the granularity of the system processing. If we have a hang of 12 boxes that are all driven from the same processor output, we can set the level and EQ for the whole array, but we don't have the ability to address any variance within the coverage of the array. If we can divide the drive of the array into smaller zones, we will be able to make some additional improvements to the high frequency uniformity by adjusting high frequency EQ per zone, a practice known as "shading".

Recall from our discussions of directivity that in the higher end of the spectrum (let's consider above 1 kHz as a rough cutoff) we can get meaningful results by operating on individual elements or small groups of elements with our EQ decisions, while at low frequency we need to address the array as a whole due to the overlap. Generally, we can expect to cut back some of the high frequencies on the bottom boxes that are closest to the listeners, and boost some of the high frequencies on the top boxes that need to throw the farthest and will suffer the most from air absorption.

The good news is that most modern systems operate with either 1 or 2 elements per processor output or amplification channel, although you may still encounter 3 or 4 box zones when an effort is being made to save money. This type of consolidation limits our ability to make finer adjustments at high frequency, and also the ability for manufacturers' proprietary beam steering and air compensation algorithms to do their work. More zones are always preferable, and two boxes per processing output is a reasonable compromise between system cost/complexity and the ability to create a high degree of uniformity.

Processing Structure

At its most basic, a signal processor can be thought of in three distinct conceptual segments: Input processing, routing / matrix, and output processing.

Signals come in from the mixing console(s) into our processor's inputs, where we have level control to allow the mix engineer to achieve their target SPL in the room while still leaving their main faders and meters hitting at a comfortable level as per their preference. Some mix engineers may be content to adjust their master or matrix fader levels to achieve the target SPL in the room. Others may wish to keep their output levels from the console at unity and ask us to adjust the input gains on our processing until the system is achieving the desired SPL.

The EQ on the inputs is our overall tonal control of the system, and this is where we will make tonality adjustments requested by the mix engineer.

From there, the board mix (which might be sent to us as L/R, L/R/Sub, or L/R/Sub/Fill depending on the circumstance) gets routed to the various processor outputs as desired. Left to MAIN LEFT, Right to MAIN RIGHT, and so forth. I use ALL CAPS naming convention to indicate outputs from my processor to the different system zones. If they are sending us a separate sub feed, we'll usually send it to the SUB at unity.

If the mix engineer is not sending a dedicated subwoofer feed, we'll probably sum Left and Right into the subs at -6 dB. Large format systems with subs flown alongside the mains are sometimes treated as a large L/R, in which case L would go to the left sub hang and R to the right.

Same with front fills: if we have a fill feed, send it on its way. If we have Left and Right, we'll probably sum them together at -6 dB into the front fills. I will sometimes take just one side if the mix engineer is using a lot of time-based effects to create a sensation of width and we wish to avoid the resulting comb filter if both sides were summed to mono. This is a decision best made in cooperation with the mix engineer.

Side hangs are typically "flipped" - so Main L goes to the Right side hang, and Main R goes to the Left side hang. This way, listeners who are in the coverage of multiple sources - which is a lot of people in an arena, for example - will still benefit from some sensation of width, and any combing interaction between the two sources in the overlap area is less objectionable. Otherwise, only the folks seated near the center of the venue between L and R that benefit from the width of the mix. This simple routing change has virtually no drawbacks and can improve the listening experience for many listeners, so we should do it.

Remember that this only has a benefit for listeners who are in the coverage of more than one part of the system. If you have a split center cluster consisting of a loudspeaker covering the left side of the house, and a loudspeaker covering the right side of the house, listeners are only experiencing one or the other, so this is not a stereo system and should not be driven as such. The same would be true for a very large format design such as typically found in a stadium, where most listeners are only in the coverage of a single source.

From there, we can control the level, polarity, EQ and delay to each output as needed throughout our tuning process.

Three-Tiered Processing Approach

I employ a three-tiered approach to system processing, which allows me to work through the system as quickly and as efficiently as possible.

Tier One is the overall system. I can set level and tonality for the entire show. This is typically done at the processor inputs that are receiving the board mix, and how I respond quickly to requests for tonal changes from the mix engineer.

Tier Two is each individual subsystem: gain/polarity/delay/EQ for each hang or major system zone. Main Left, Main Right, Side Left, Side Right, Front Fills, Subs. This is the major building block layer of the system tuning, where I can easily balance all the elements of the system.

Tier Three is each individual loudspeaker, or pair of loudspeakers if it's two per processing output. This is where I implement high frequency shading on the arrays and per-box timing on the front fills as needed.

Accomplishing this depends a bit on how the system is laid out and driven. If we have a system that has the presets and processing built into the manufacturer's amplifiers, I will use a front end processor to take the board mix and distribute it to each of the subsystems (Tiers 1 and 2) and then do my shading and individual fill timing in the amplifiers' control software (Tier 3). If we have a completely active system with DSP built into the loudspeakers, I'll do it the same way (Tier 1 and 2 in the front end DSP, and shading in the control software for the loudspeakers directly).

If there is no DSP downstream of the main processor (hitting "dumb" amps or feeding powered loudspeakers with no user-facing DSP control) then we might have to do double-duty and use the output processing on the DSP both for applying EQ to multiple outputs together (entire array) and shading (individual outputs). Processors that allow multi-select make this relatively easy. Even better are processors that allow a relative change between selected outputs.

For instance, imagine we have a main hang of 8 boxes driven by 4 consecutive outputs on the processor (two elements per output). We wish to do some overall EQ, and some high frequency shading per zone. If during the show we decide that the entire hang is a little bright and we want to bump down the high frequency filters, we can select all four outputs together, and bring the filters down by a set number of dB while still retaining their individual offsets. Relative adjustment is a powerful tool to get results quickly.

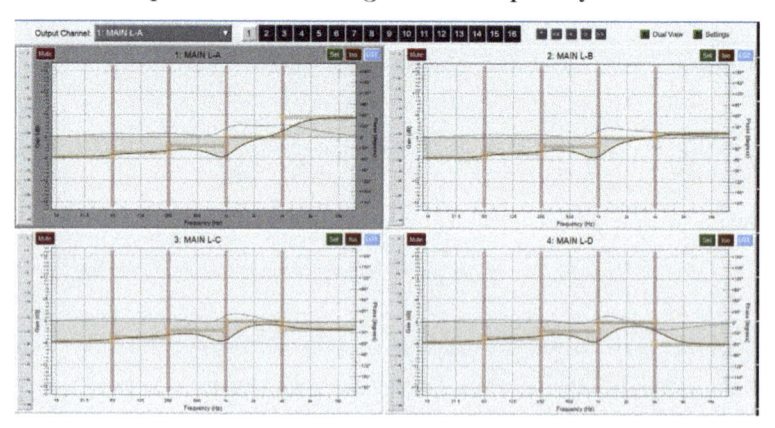

Figure 38 – Meyer Sound's Compass software for the GALAXY processor, showing EQ for Main Left zones A,B,C,D. They can be adjusted as a group or individually. Note the overall curve is the same, but with different HF shading. (Compass screenshot courtesy of Meyer Sound)

Other processors handle this via layers and "overlay" EQs, which allow a user to logically group various outputs, add a filter to the entire group, and still operate EQ on individual outputs as well.

If you're using a processor that offers the ability to route an output zone directly to another output zone (like bus-to-bus routing on a mixing console) then this becomes trivial to accomplish – a "PA LEFT" zone can be routed to three individual output zones for L-A, L-B and L-C, which then drive individual zones of the array. Adjust the entire array's EQ via PA LEFT, and do the high frequency shading on the individual zones.

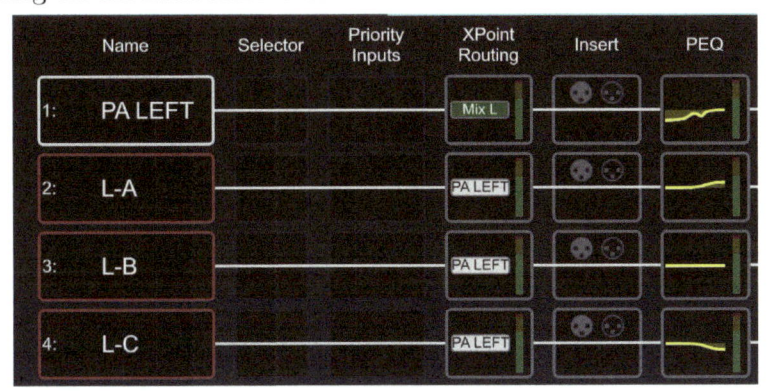

Figure 39 - Allen & Heath AHM System Manager software for the AHM64 processor, showing PA LEFT zone with overall tonality EQ, routed to three sub-zones A,B,C with HF shading (AHM System Manager screenshot courtesy of Allen & Heath)

"Freewire" processors allow the user to configure processing modules in any desired order and arrangement, so it's easy to insert multiple stacked equalizer and level control blocks as needed. Freewire processors are less common in the live events world but are often encountered in install situations.

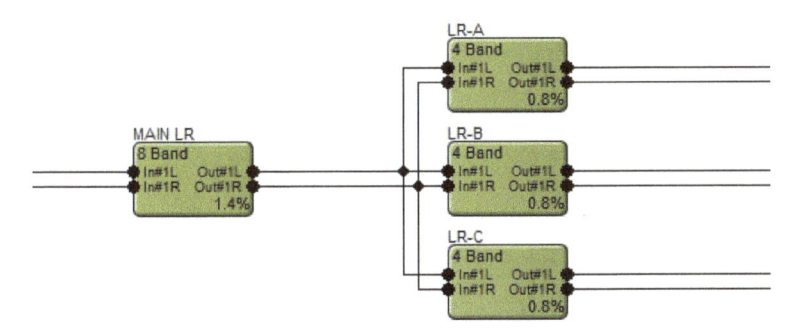

Figure 40 - Symetrix Composer software for the Prism processor, showing cascaded EQ stages for the entire array and then individual HF shading per zone. (Composer screenshot courtesy of Symetrix)

BETWEEN THE LINES /115

Signal Generator Routing

Let's spend a moment to consider the options for routing our signal generator from our analyzer into the system. With a realtime dual channel audio analyzer, we can use any appropriately full-spectrum test signal to measure the system. It doesn't necessarily need to be the generator built into the analyzer. However, there are some benefits to doing so - ease of control and some additional technical and mathematical benefits that are best left to a Smaart® training course.

Assuming we want to route the signal generator from our audio analyzer into the system, what's the best way to do that? The "signal path of least resistance" is to simply plug into a free input on the mixing console and use the mixing console to route the signal where it needs to go. Unfortunately, the "signal path of least resistance" is also the "signal path most likely to have undesirable and unnoticed processing in it."

Don't be the person who tunes their entire system with a high pass filter, multiband compressor or bus EQ on the reference signal because you didn't realize what was happening inside the console. This also assumes that we have access to the console and can freely change the routing, which is often not the case.

Best to leave the console to its own devices and inject the signal directly into the processor. Give it its own input and route it where it needs to go, and then the mixing console can be connected at leisure, and we don't need to spend time messing with the processing on the FOH console.

Additionally, being completely separated from the FOH console means that I can operate independently and do my work regardless of what the FOH engineer is doing. Maybe the console isn't up and running yet because we are waiting for a stage rack to power up, or the FOH engineer needs to use the console to work on the mix using virtual soundcheck. Injecting our generator into the system at the processor inputs means we can work uninterrupted, and so can the mix engineer.

Generator Gain Structure

One more hidden consideration here: we are probably (hopefully) going to tune our system at 20 dB or so below show SPL, so we need a different gain structure on the generator input into the DSP than we would for the board feed which will drive the PA to full show SPL.

It might be tempting to plug your interface's line-level output into the system processor, and then turn down the analyzer's generator level to -36 and start your work there, but this is risky. If you bump a cable and your interface comes unplugged, or your computer goes to sleep, you're going to put a very loud, scary and potentially damaging POP through the system at full scale. This is also a risk if you mis-type the generator level and it comes on at 0 dB full output.

A better, safer, and far more "cover your ass" solution is to run your signal generator closer to full output, perhaps -12dB, and then attenuate the generator input at the inputs to the DSP. That way you've taken a solid 20 dB off the "scary factor" of the resulting POP if there's an interface hiccup or you have a ham-fisted typing moment, while leaving yourself some room to turn the generator up or down within the analyzer without affecting the measurement.

Learn to operate your DSP with fluency so you can spend time focused on more important things.

CHAPTER 16
The Four Principles of Alignment

Now it's time to verify and tune the system. A conceptual note before we start: People often ask for a list of steps or a "checklist" to approach tuning a system. Every system is different, and so is every use case, so that simply won't work. What I offer instead is a mental structure, mindset and school of thought that you can use to *generate a list of steps* to tune each and every system you encounter, based on four simple "first principles."

Before we commence tuning, we need to do something far more important: verify our system. Make sure it's aimed correctly, wired correctly, all the loudspeakers, amplifiers and processors are connected and powered on, make sure that anything connected to a network is identified and enumerated properly, make sure control software can talk to amplifiers and processors, etc.

It is truly astonishing how often I am asked to come in and tune a system only to discover that Main Left and Main Right don't match. Something is wrong, broken, miswired... stop and find the issue before you start tuning, otherwise you're going to have to go back and do it again. Don't wallpaper over a huge hole in the wall. Fix the hole first.

I will route the signal generator to each element of the system - starting with entire subsystems (Main Left, Main R - do they sound the same, same tonality, same level? If not, stop and fix it). Side left, Side right. Sub left, sub right. And on we go until I've heard everything and confirmed that Left and Right are matched in level and tonality, not reversed, etc.

Then I go through the system zone by zone or box by box. If one box in a hang sounds quieter than the others, stop and find out why. If one box sounds tonally lighter or darker than the others, stop and find out why. Don't try to plow forward and try to compensate for a bad cable or a blown-out driver with gain or EQ.

Quite often, the verification process reveals a problem. It might be relatively minor - I encountered a system in which the four front fills differed in level over a range of about 3 dB - or it might be severe - I was recently handed a system where Main L and R were reversed, 5 boxes didn't work on the left hang, the front fill feed came out of the bottom box on the right hang, and the mains didn't have the same trim heights or site angles. Don't worry, I'm sure we can fix all of that with EQ.

Once everything is verified and the system is confirmed to be working properly, the tuning can begin. Since we've just verified that the system is symmetrical (you *did* verify that the system is symmetrical, right?) we can tune one half and then do a final check at the end to make sure we are still symmetrical.

The Four Principles

As simplistic as these may seem, these four principles will guide you through any system tuning process.

> 1. **Each source has an assigned area of custody, within which it is responsible for level and tonality (gain and EQ).**

We sorted out the "custody" part in the design stage. Now during the tuning, we will use the analyzer to set the desired level and tonality within that custody area. This means adjusting gain and EQ.

> 2. **Once level and tonality are set for two adjacent systems, find their equal-level seam and make timing decisions there.**

Since two sources will interact most where they are equal in level, the seams are where the timing matters most. As we move in either direction from that seam, one source gains level dominance and summation effects decrease in severity.

This means adjusting delay to control the relative timing at the seam. We must do this *after* we've made our level decisions, because as we adjust relative level between different sources, we will also be changing the location at which they meet at equal level.

> 3. **Begin with systems that have the largest custody areas, which establish the overall level and tonality. Work your way down. Supporting systems restore the target level and tonality in their custody areas.**

The "papa bear" system is the Mains. It covers the most people.

Measure it in its coverage area and we'll have an idea of the starting point for the level and tonality. We'll adjust the tonality until we're at the desired target curve. The mains and subs work together cover most of the listeners and establish the target curve that the rest of the system will support.

Once we are happy with that, we'll work our way down through the supporting systems. We will always choose the next system to work on in order of decreasing custody size. If at any point we get interrupted or run out of time, we've spent our tuning time working on the systems that cover the most people and have left the smallest possible number of seats unaddressed.

Supporting systems restore level and tonality within their custody area. Once we know what the mains are doing, it's easy to measure the next system down the list in its own custody area and see how it compares - any necessary level or tonality changes will become obvious.

4. Sources consisting of multiple elements are assembled by addressing uniformity across elements, and then addressing the combined behavior.

This is just a re-application of Principles 1 and 2. When dealing with a combined or "composite" source such as a line array, we can address variance from front to back in the frequency range where the individual elements have some independence of custody (high frequencies), and once we've shaded to achieve front to back consistency at high frequencies, we can set level and EQ on the entire array to hit our level and tonality target. We can then consider this to be "one source" and continue to combine it with the next in line (such as a subwoofer). In practice, these two operations (shading and overall tonality) might happen at the same time, especially with multi-microphone workflows, but conceptually we are accomplishing two distinct tasks.

Any time you are not sure where you should start, or what you should do next, consult these four concepts and the way forward will become clear.

Choosing Mic Positions

Every time you place your measurement microphone, it should be having an existential crisis ("What am I doing here? What is my purpose?"). If you can't answer that question, stop and figure out a good answer before you continue. Measurements provide answers to questions. Ask yourself what you're trying to learn, and then ask yourself where the mic needs to go to get that answer.

If you're working gain or EQ (**Principle 1**) you're making decisions about what the loudspeaker is doing within its custody area, so that's where the mic should go. A few measurements taken from various positions within the bulk of the custody area can provide a "consensus view" if you're worried about getting distracted by localized micro-ripple. Applying some smoothing, and averaging together multiple measurements can help to illustrate the trend as well.

If you're working on timing at a seam between two sub-systems (**Principle 2**) then you're asking who's arriving first at a very specific location: wherever the two sources are equal in level, so that's where the mic should go.

This equal-level point is usually quite straightforward to find by ear. To confirm with the analyzer – measure each source by itself and look for the high frequency portion of the traces to line up. If they're not equal, move the mic closer to the source that's lower in level and try again.

Sometimes the "seam" is a larger area, such as the overlap that might occur between flown and ground-based subwoofers, or between mains and delay towers at a festival. The same logic still applies, but as the area of the overlap increases, we have to think more carefully about where it makes sense to make our timing decision, and must be honest with ourselves about the extent to which that timing will hold throughout the overlap region.

We are inherently taking a reductionist view here – the venue might have 1000 seats, but we aren't going to measure all 1000 seats to get the information we need to tune the system. So how many positions do we need? The answer is "just enough to answer all our questions, and no more."

My approach is to optimize my speed and efficiency by *gathering data until I reach the point that any additional data doesn't change the decisions I'm making*. At this point, gathering any additional data simply decreases efficiency. The Four Principles help us make "targeted strikes" into our system's coverage by understanding what's happening doing at key locations. If we understand what a system is doing in its custody area (**Principle 1**), and we understand how two systems are interacting at the seams (**Principle 2**), we know what's happening in between.

This is the key to achieving both speed and precision – we are eliminating unnecessary steps, not cutting corners. We are also eliminating any time spent idle, thinking about what the next step is. The Four Principles lay the path before us as long as we clearly reasoned through our game plan before we started.

When I put my mic down, I know exactly what information I'm seeking at that location, what decisions I'm going to make because of that information, and where the mic is going next. There's no "dead air."

Array Zoning

It might seem daunting - and not particularly expedient - to approach high frequency shading on a per-box level with 8, or 12, or 16 elements in an array. Does a 16-box array mean we need 16 mic positions? No. My approach is to form logical zones based on the audience geometry we're covering with the array. The section view of the prediction is very useful when making this determination.

If an arena design has 4 boxes aiming into the 200 level, 5 aiming at the 100 levels, and 6 on the floor (three mains and three downfill boxes), I might choose to address those as four conceptual groups in the process, identified with the letters A, B, C and D starting from the top down.

Typically speaking, we can expect to end up with larger box counts with narrow splays in the upper zones, and smaller box counts with wider splays in the lower zones, which makes sense considering the "angular rate of change" as we move down the array.

Let's say that "A" is the 4 boxes into the 200 level of seats, "B" is the 5 boxes covering 100 level, and we'll divide the 6 on the floor into "C," three boxes covering the rear half and "D," three in the front half.

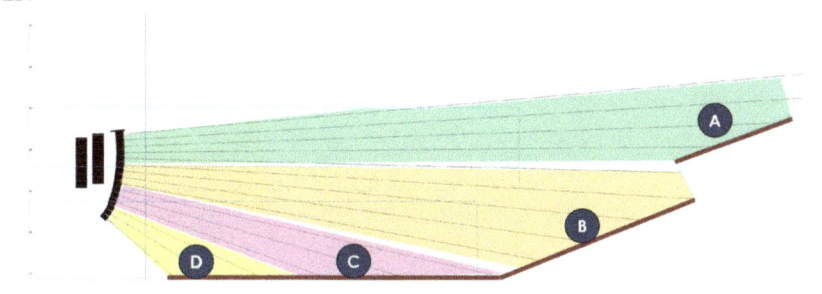

Figure 41 - Array zoning strategy for a 15-element array based on audience architecture

Good news - now we know exactly where our mic positions go, and

BETWEEN THE LINES /125

also where to go in the DSP to approach the high frequency shading. It's just a matter of measuring the array from each position, comparing the resulting high frequency responses and adjusting the shading in each zone until the variance is reduced as far as we decide it can (or should) be.

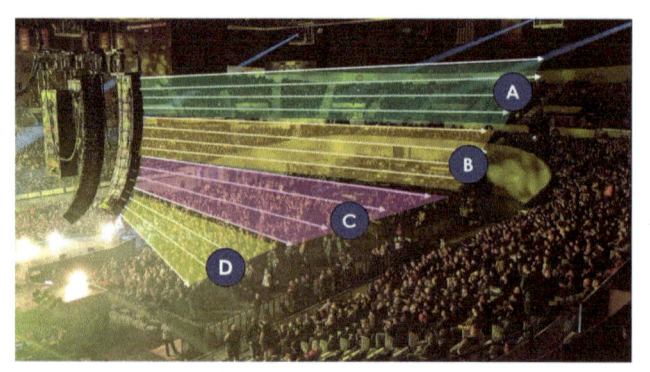

Figure 42 – An example of array zoning in practice

High Frequency Shading

Recall our earlier discussion about why it might not be a great idea to use high frequency shading to completely eliminate all high frequency variance. Another factor to consider is that the humidity will probably rise when the venue fills up with humans, but management says we have to let them in anyway. Some of the high frequency lost to air absorption will be restored, particularly in the top zone which is throwing the farthest.

If we brought the high frequency response in our top zones completely back to our target response with shading, and then the humidity increases, air absorption is reduced and the system now sounds unpleasantly bright in the back of the room. In larger indoor venues, I typically will allow the high frequency to fall a few dB short in the furthest zone and let the environmental change tip it into the sweet spot. Of course, I will always take a walk up there during the show to listen and adjust as necessary.

If we've shaded in four zones, we might have the high frequency EQ set to +3 dB in Zone A, +2 dB in Zone B, -1 dB in Zone C and -4 dB in Zone D. This means that between boxes 9 and 10 (the transition from Zone B to C) we jump by 3 dB in the HF, and then again between boxes 12 and 13, we have another 3 dB jump. We can smooth this transition a bit and make it more audibly transparent by taking advantage of our processing granularity.

Once the overall high frequency zone shading has been established, we can go back through and feather the edges a bit to spread the filter gain differences over a few elements.

For the above example, our high frequency filter gain shading per box started off at:

+3, +3, +3, +3, +2, +2, +2, +2, +2, -1, -1, -1, -4, -4, -4.

After some feathering of the filters, we might end up with the following filter gains:

+3, +3, +3, +2.5, +2, +2, +2, +1.5, +1, 0, -1, -2, -3, -4, -4.

This approach can make a real improvement in the zone transition regions and is especially useful to keep the system response seamless around mix position, where tonal decisions are being made.

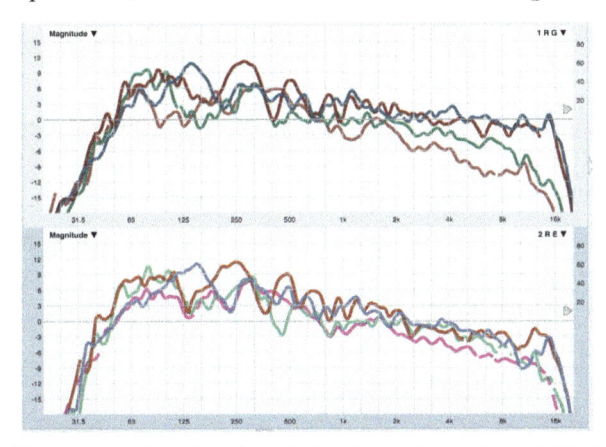

Figure 43 - Measurements taken front-to-back of a large format 14-element array, before and after high frequency shading

What About Gain Shading?

Gain shading is sometimes used as a substitute for doing the shading with high frequency EQ filters. It's less optimal if the goal is simply to smooth out the high frequencies, in that a gain change affects the loudspeaker over its entire frequency range. As we know, the entire array is playing together at low frequencies, so reducing the gain on elements reduces low frequency headroom for the whole system. Gain shading also has an effect similar to shortening the line length, since turning down the bottom boxes reduces their energy contribution and therefore their contribution to the directivty, with the extreme case being turning them off completely, producing the same effect as not hanging them at all.

However, this is another example of the Magnitude Fallacy, as the amount of lost low frequency headroom is less than one might expect with typical amounts of shading. A 12-box array with 3 boxes shaded down by 3 dB and 3 boxes shaded down by 5 dB loses 1.8 dB of total low frequency headroom. However, gain shading can also be disruptive to manufacturer-implemented beamsteering algorithms and so should be used with caution in such circumstances. Note that this isn't a reason to avoid it altogether - gain shading can be a helpful approach, especially when we don't have adequate DSP resolution to do our shading with EQ. When in doubt, we have wonderful tools (prediction and measurement) that allow us to evaluate the effects and decide whether we wish to implement the change.

I'm not scared to get my hands dirty and do shading -whether that be high frequency shelving EQ or gains - if it gives me the uniformity improvements I'm seeking.

If you're still fearing the phase gremlins with regards to the high frequency EQ, don't.

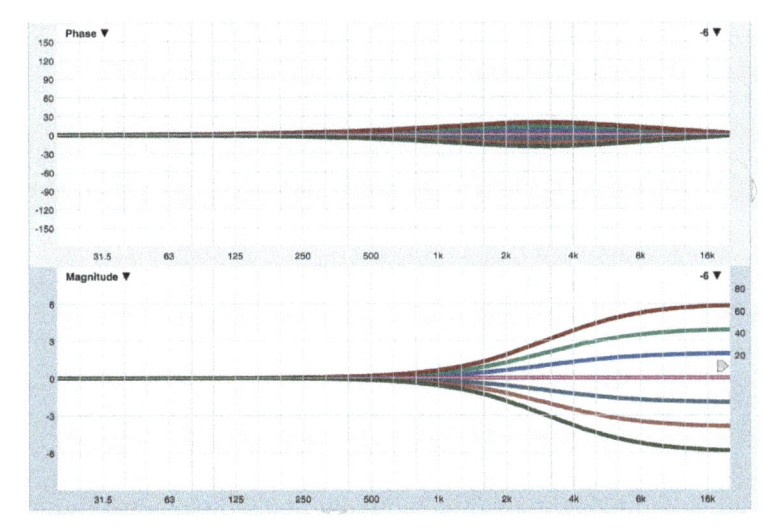

Figure 44 - Magnitude and Phase response of HF shelving filters used for shading (Smaart® v8)

The graphic demonstrates the phase response of high shelving filters over a gain range of +/- 6 dB. Vector summation math tells us that if our phase offsets are within 55°, we will remain within 1 dB of an ideal summation. There's nothing here to be alarmed about.

Fuzz at the Bottom

Down at the bottom of the array, we have another issue to be aware of. When we're standing in the coverage of the bottom boxes of a curved array, a decent portion of the high frequency energy we are hearing is coming from the side lobes radiating from the boxes further up in the array. This is readily visible in the LiveIR measurement taken from such a location.

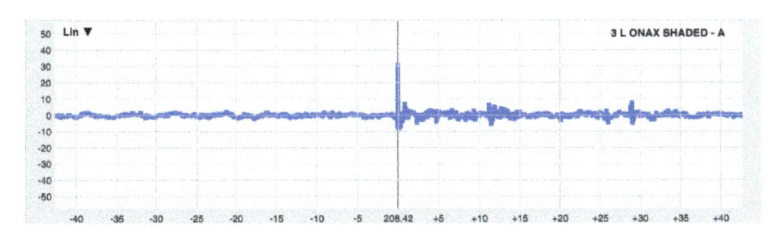

Figure 45 - LiveIR of a large-format system at ONAX A (Smaart® v8)

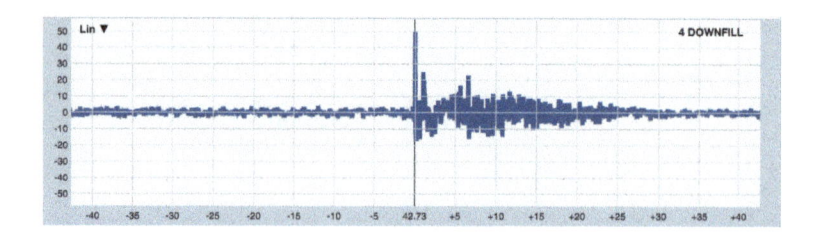

Figure 46 - LiveIR of the same system taken ONAX to the bottom 3 elements. Note the "fuzz" of late arriving energy from boxes higher up in the array. (Smaart® v8)

It's probably going to be a bit bright down front, and we'll want to do some shading, but turning down the high frequencies on those bottom elements will only work up to a point – because a lot of that energy doesn't belong to those boxes, but is coming from further up in the array.

As we continue to shade the bottom boxes, less and less of that energy is direct, and an increasing amount of it is coming from elements higher up and won't respond to shading. Past this point, we are not improving the tonality at all, but we are making it sound "further away" and less clear, and can create a weird imaging situation for the listeners down there because it may seem like those elements are not on.

Find the point where the shading loses effectiveness and then bring it back up a dB or two. Listening will confirm whether you've got enough direct sound to avoid strangeness. A little bit of brightness is not the end of the world, we just want to control it as much as we can. Like everything else, we are seeking the best compromise.

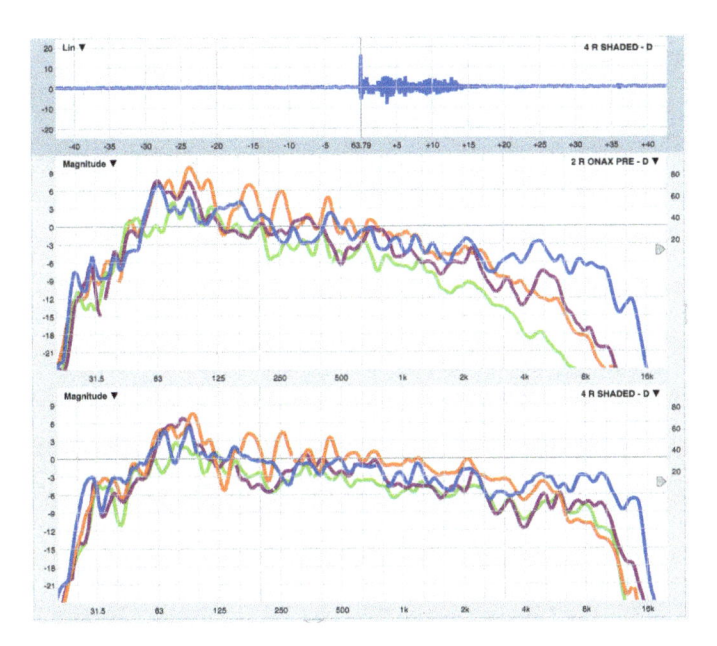

Figure 47 - Front to back measurements before and after high frequency shading on a large-format array. The excess high frequency energy at the bottom (blue trace) is reduced but not eliminated. Notice the "fuzz" in the Live IR from that location. (Smaart® v8)

Multiple Mic Workflows

Such shading operations can be much quicker work if we have multiple mics to deploy at each of our on-axis zones, rather than moving one mic between them over and over.

With a multi-mic workflow, the first step should always be a mic verification - place all the mics together, turn the system on, make sure they all give you the same answer. Adjust preamp gains as necessary until all the traces overlay. This creates a valid basis to make comparative decisions between the mics.

BETWEEN THE LINES /131

Ground Plane / Boundary Mic Positions

Discuss system measurement for long enough and someone will suggest a "ground plane" or "boundary" mic position, whereby the measurement microphone is placed on the floor in order to eliminate the floor bounce from the measurement.

This works as advertised, but we must be mindful of what we are trying to accomplish. Boundary effects being what they are, the tonality you measure at floor level doesn't necessarily represent the tonality experienced at ear height. Exercise extreme caution when using the ground plane position for tonality (EQ) decisions, because no one listens to the show at ground level.

The boundary or ground plane mic position can be helpful for making timing decisions, particularly if we are interested in the low-mid region of the response, around 200 Hz where the ground bounce tends to manifest. If you need to - for instance - measure main-sub timing, the ground plane mic position can produce data that's easier to interpret.

It's also a valid position for comparative measurements (multiple array zones on an arena floor or outdoors, for example). Since we're looking for variance, we can see it easily by comparing multiple measurements taken at floor level, even though the overall tonality might not be representative of what's happening at ear height.

One important application of the ground plane position is verification of individual loudspeakers before an install, for example. The photo below shows a microphone affixed to a stage floor. One array cabinet at a time is being measured and compared against the others before they are installed in the venue, which allows straightforward verification that all the elements match, a task which becomes far more difficult once the rig is in the air.

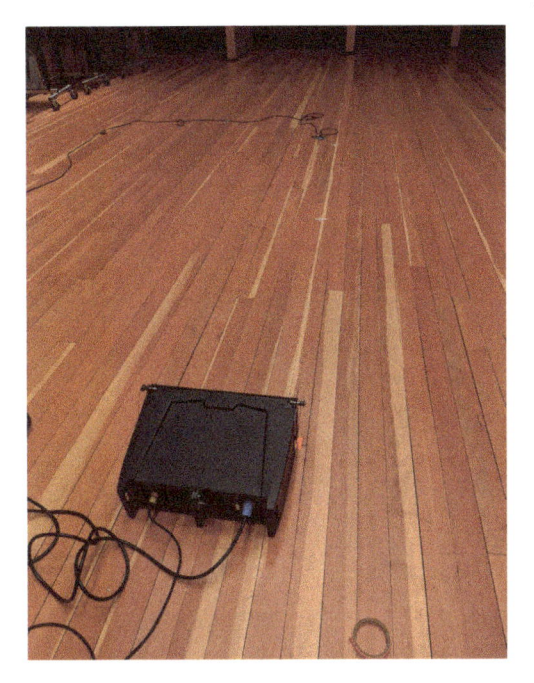

Figure 48 - A new line array element getting a "checkup" with a boundary mic position

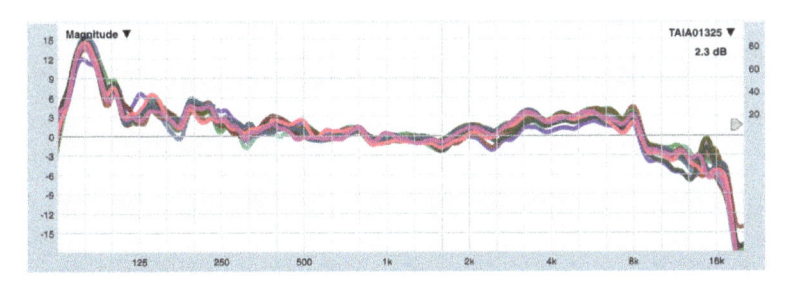

Figure 49 - Traces from 14 new line array elements taken from the same boundary mic position to verify that all is working properly (Smaart® v8)

Personally speaking, in practice the ground bounce is rarely an issue for me - it's easily identifiable as such (it shows clearly on the Live IR as a second arrival) and it's easy enough to "see through it" once you know it's there. As such, I very rarely use a ground plane mic position in practice during a tuning session.

If the floor bounce is problematic, you can also gather several measurements with the mic at various heights and positions around the same area, then average them together, which can suppress or eliminate the effect of the ground bounce while remaining more tonally representative.

"I have never been in a venue that didn't have a floor." - Jamie Anderson

---

SECTION 6
Strategy

CHAPTER 17
Tuning Strategy

Here the rubber hits the road. This section steps through my order of operations and decision chains for a series of alignment tasks representative of various venues that generally increase in size and complexity as we go. It is not a set of instructions, and hopefully by this point in the book you have retained the concept that there is no "one size fits all" in this work. Rather, you are encouraged to think about the "macro" here - how the same mindset and workflow philosophy yields an efficient and fruitful process in each case. Can you spot how the same "first principles" inform choices in each case? Here they are again:

1. **Each source has an assigned area of custody, within which it is responsible for level and tonality (gain and EQ).**

2. **Once level and tonality are set for two adjacent systems, find their equal-level seam and make timing decisions there.**

3. **Begin with the systems having the largest custody areas,**

which establish the overall level and tonality. Work your way down. Supporting systems work to restore level and tonality within their custody area.

4. **Sources consisting of multiple elements are assembled by addressing uniformity across elements, and then addressing the combined behavior.**

If you've been following along up to this point, these four simple bullet points should be all you need to understand why I'm doing what I'm doing at each step along the way. The goal here is finding the thread - looking at the system laid out before you and assembling a mental order of operations that will bring it into uniformity. The following examples should read as variations on a theme, as I work through the same process over and over again.

When I approach a tuning, I start with a mental inventory of subsystems, ranked in order of how many listeners they cover. Then I start working my way down the list following the four principles listed above.

I've begun each example with a brief outline of the system design (what we have and where it is) so you can think about finding your thread before reading about mine. With this knowledge in hand, you can build an effective approach to any system you're asked to tune, regardless of scope or scale.

Studio L/R Nearfield Monitors

The simple left-right pair of nearfields is the most straightforward of alignment tasks. Our listening area is the size of a single listener, and we know exactly where they are going to be, so we can completely sidestep any considerations of variance over a larger space (don't worry, we'll return to that soon enough). This also means we can be a little more aggressive with our equalization strategy, as we can pay less mind to any undesirable effects that might happen off axis. We focus our efforts on improving the situation as much as possible at the single listening position where mixes are evaluated, and where artistic decisions are made.

Verification is a large part of the job here. With our microphone at the listening position, does Left match Right? You might be surprised at how often the answer is "no," even with relatively expensive loudspeakers. We are looking for matches in level, tonality, polarity, and arrival time. If one side is arriving early, it could easily be fixed with delay, however it would be prudent to investigate why that might be the case. Is the configuration asymmetrical? Does the speaker placement need to be adjusted?

If there is a subwoofer present, it is easy to measure its response from the listening position, and observe the relative phase response compared to that of left and right (which should match). A single sub needs to sum with Left and Right playing together, so at this point, it's best to measure with both L and R on, save the combined trace, and use that to evaluate the relationship with the subwoofer. Some comb filtering in the high frequency will be observed if the arrival times are not exactly matched, but that is unimportant at this point, as we're focusing on the low frequencies – specifically the frequency range over which both the sub and the mains are contributing at similar level.

Either the L/R or the Sub will get delayed – whomever is arriving first – until they are phase-aligned throughout that frequency range, then you may want to measure again with all three sources on to confirm the summation.

The combined system can then be equalized to any desired tonality, which for studio work is often a gentle downward tilt, or a relatively flat response with a slight low frequency rise.

In studio environments, there is often a monitor controller that switches between multiple sets of loudspeakers. It's advisable to ensure that the relative level is matched between each system. These devices also sometimes contribute insidious polarity inversions and L/R level offsets, so don't trust them as far as you can throw them.

Coffee House Point Source L/R/Sub

Now we will leave the brooding, creative studio environment and go out in public to listen to brooding, creative music with other humans. Local favorite Acoustic Terry is performing solo tonight at the Mojo Room. This is a small coffee club with point source "speakers on sticks" on either side of the stage, and a sub beneath each speaker pole.

Now that we have a listening area that includes more than one human, we no longer have the luxury of ignoring the concept of variance. It's back in full force, but since we have speakers on sticks, our ability to tame it is inherently limited. The best thing we can do is get those sticks as high as possible. This reduces our range ratio right out of the gate.

Raising a loudspeaker up doesn't significantly impact the distance it needs to throw to reach the furthest listener, but it can dramatically increase the distance to the closest listener, which reduces the range ratio, and gets the folks down front out of the "hot zone" where the distance doublings happen very quickly, and where they are at increased risk of a sound exposure hazard. Indeed, one of the most effective ways to reduce level variance in virtually all situations is to get the system off the ground.

Since the loudspeakers have their maximum output on axis, we want to aim them directly at the furthest listener. From that point, moving forward across the audience area also means walking off-axis of the horn (provided, of course, that the loudspeakers are above head height), so we offset the axial loss with the proximity gain to a degree, at least in the high frequencies. This mechanism runs out of steam after we get to the edge of the horn (6 dB down at high frequency) and then we either need a different loudspeaker with a different vertical dispersion pattern, or we need to call in a fill speaker for reinforcement and the game begins again. Since it's a coffee house gig and they probably don't have Front Fills as a budget item, we must live and let live.

0. Verify
1. Main: Level + EQ
2. Sub: Level + EQ
3. Main + Sub: timing, combined EQ

Our roadmap begins, as it always should, with verification - "spray" left, then right, then each subwoofer with pink noise. Make sure left and right sound the same, and at the same level. Powered speakers are low-hanging fruit for level knobs being inadvertently adjusted. If any of these boxes have built-in DSP, let's make sure the settings match and both subs are in the same polarity.

Tuning begins on axis with one side of the mains - I typically start with whichever side offers a shorter path from its custody area back to my computer. It doesn't matter much in a coffee house, but in an arena or festival setting with a large FOH area behind barricade, and only one place to enter, it will save us some steps throughout the process, so let's get in the habit of evaluating that before we begin a tuning.

Although we must settle on one EQ for the entire coverage area of this loudspeaker, we might still want to take a few measurements front to back so we can get an idea of what the variance is like and make a "best compromise" EQ decision. A gently tilted tonality is typically appropriate here.

Then let's turn on the corresponding subwoofer and adjust its level relative to the main to create the most useful extension of that gentle tilt. If it will help you feel less brooding and more creative to do so, pick an "average" depth, perhaps halfway back in the listening area, and observe the relative phase responses over the frequency range where the main and the sub overlap, and delay whomever is earlier until the phase traces overlay.

Capture the combined response and check it against a symmetrical position on the other side. Then fire up your reference music and do some critical listening, adjusting the overall EQ as needed.

Music Club L/R/Sub/Fill

Acoustic Terry has gained a following and has managed to book a performance at the local music club, the Stumbling Donkey. Now we have a properly flown main system consisting of a pair of point sources boxes per side: a Main box that covers most of the listening area (a flat floor) and then a smaller downfill box that covers the first third of the listening area.

On the floor are four single subwoofers spaced evenly across the front of the stage, and four small frontfill boxes across the front edge of the stage as well. Mix position is, rather predictably, a slightly elevated and structurally questionable wooden platform in the rear of the venue, so we'll need to measure from back there once we're done tuning so we know how representative (or not) the mix position is compared to the rest of the audience area.

First we verify the rig: spray Left Main, Right Main, Left downfill, Right downfill (I do them in pairs like this because it's easier to hear when there's a lack of symmetry), and then we want to verify the subs and fills.

However, all four subs share a single processor output, so the only way to hear them one at a time would be either to shut off amplifiers or unplug the boxes. Since a club is a "higher risk environment" when it comes to damaged / mismatched / mis-wired / beer-filled loudspeakers, I'll take the extra time to unplug three subs and check them one at a time.

Same thing with the front fills, but the fact that they're full range and have a more controlled directivity saves us some work – spray all four, walk down the line and make sure they all sound the same (same tonality, same level). A blown driver or gain difference is easy to hear in such a case, as is a polarity inversion if we listen at the seams where two fills overlap.

The Mains consist of two elements per side, so we turn to our Four Principles to guide us: we'll start with the majority shareholder main box, establish its response and level, then make the downfill match that response in its custody area.

Then make those two sources work well together via timing and combined EQ, after which we'll consider the combined source the "Main" and mate it with the subs. Main + Sub together is what establishes the overall tonal response of the system, which the front fills will then extend into their custody area.

0. Verify
1. Main: Level + EQ @ onax
2. Downfill: match Level + EQ @ onax
3. Main + Downfill = MAIN (timing, combined EQ) @seam
4. Sub: Level + EQ
5. MAIN + Sub: timing, combined EQ, establishes target for rest of rig
6. Front Fill: define custody (depth and frequency range)
7. Front Fill: Level + EQ @ onax
8. MAIN + Sub + Front fill (timing) @seam

Starting with the left side, the Main Left loudspeaker covers the most people so let's fire it up and measure it on axis around head height. EQ as necessary to achieve the target response – I'm going with a gentle downward tilt again – and once I have a smooth desirable response, turn off the main, and turn on the downfill.

Let's measure from onaxis to the downfill and set the level and EQ to accomplish the same tonality, but I want it to be 1 dB higher in level, since it's closer to the stage and the accompanying stage wash. We can nail this by adding a -1 dB trace offset to the magnitude trace in our analyzer, then adjusting the level of the downfill until the trace overlays perfectly with the mains. It's a smaller box so let's make sure it doesn't have any excess high frequency extension that we won't be hearing further back, then remember to clear the level offset when we're done ([Y] hotkey in Smaart®).

Now that we have Main and Downfill achieving the desired level and response in their areas of custody, we fire them both up and find the equal-level seam. It is easy to find by ear with a little practice - turn them both on and walk through the seam until you hear the strongest comb filtering interaction.

You can also find this via analyzer if you're having trouble, by measuring one source at a time to see whether the high frequency is at the same level for both. If one source is still higher in level at that location, move the mic towards the other system and try again until they show up the same - then you've found the seam. Bob your head back and forth through this seam so you can train yourself to recognize it by ear next time.

Whomever is arriving first here gets delayed. I turn on the main, set the analyzer's measurement delay to that arrival, then capture the measurement and turn off the mains. Turn on the downfill. If it's early, it can get delayed back until its phase trace lands on top of the captured Mains trace. If it's late, capture its response, then switch back to the mains and delay the mains until the phase trace overlays with the stored downfill trace.

When we turn both the mains and the downfill on together, they combine into our new "Main Left". We will observe a low-frequency buildup, so we might wish to use some lo-mid EQ on the combined cluster to reduce it back to our target curve.

Once happy with that, we'll bring on the subwoofers. If they're all properly matched (you did check, didn't you?) we should have a pretty straightforward time adjusting the overall level until it gives us the desired extension to fill out the low end of our target curve. Remember that since we're looking at half the mains, we want to use only half the subs in order to keep things in parity tonally. The amount by which bringing on the full system will change the target tonality depends heavily on the amount of overlap between Left and Right mains as well as the coverage shape of the subwoofer array, so we'll be sure to adjust that by ear when we walk and listen to our reference music.

In order to time mains to subs, we have to choose where we want the systems to be in time. Center of the room tends to be a poor choice, because we're creating a single location where all the sources will pile up and sum coherently, which will happen nowhere else.

We'd rather spread the error over the space, so a third of the way over to each side tends to even things out a little bit. However, we must remember that the gains we're talking about here are limited solely to the frequency range where the mains and subs overlap, which tends to be rather limited with systems based around smaller-format main speakers that lack low frequency extension.

The more important consideration is that MAIN + SUB together gives us our complete target tonality, which is what we need to use the front fills to restore down front. It should be straightforward to determine the area where the front fills have custody (hint: they're called front fills), so let's go down there and see how big the hole is.

Walking that area with the main system on but the front fills muted reveals the size of the coverage gap - listen for where the high frequency from the mains starts to disappear. That's where the front fills need to penetrate to. Not going to make it that far when the venue is full of people? Design issue, not tuning issue. There is no processing I'm aware of that will allow the output of a front fill to pass through a human skull, although I have seen some intoxicated concertgoers try to give it a go.

The mic can go halfway between that end-of-coverage location and the front row of listeners. I take a measurement from there with the mains and subs on but the front fills off, so I can see the frequency range over which the front fills need to restore coverage. Odds are we have enough, or more than enough, low and low-mid energy down there in front already - we're standing right near the subs, and we have the off-axis contributions of the mains and downfills.

I will re-display the target trace (Main+Sub in their custody area), turn on just the front fills, and set their tonality to achieve the target above 200 Hz or so, and high pass out anything lower. It's not needed, saves headroom, and keeps some low frequency wash off the stage. Just as with the downfills, I want my front fills to be 1 or 2 dB above target further back in the house depending on program material. I hear Acoustic Terry gets pretty wild on stage.

And again - the front fills are likely to have high frequency extension exceeding that of the mains, since they are smaller loudspeakers placed closer to the listener, so we'll need to remove anything that extends past what the mains are doing in the house. With the front fills tuned and gained to restore us to target in their custody area, now we need to set timing, and we have a decision to make.

We have four front fills, and the inner pair and outer pair have different spatial and time relationships with the mains and downfills. But we only have a single delay setting to implement because all four fills share a processor output.

Principle 2 reminds us that the mains and fills will be most interactive where they are equal in level, which means the outer pair is what we care about here, since it has the more substantial overlap with the mains.

Typically, in these situations it makes more sense to go with the outer pair for the basis of our timing decision, because the outer pair is more likely to have more overlap with the main PA, and thus a higher level of interaction at those locations where both systems overlap. The inner pair of the front fill is more likely to be taking advice from Fleetwood Mac and going its own way.

Another way of thinking about the same concept: the number of audience members who are in the coverage of both Main L and Outer Frontfill is much larger than the number of audience members who are in the coverage of both Main L and Inner Frontfill, and we want to prioritize the timing relationship between these two sources that improves the result for the largest number of people.

I put the mic there and time the front fills to be 2 ms early at that location. Typically I want anyone who's hearing both sources to localize forward, to the stage, so I'll let the front fills lead by 2 ms or so at the seam. It doesn't take much to shift the brain's localization and this avoids the potential issues associated with larger time offsets.

If I am trying to move quickly, I sometimes don't bother going through the process of using the analyzer delay finder's "delta" display to tell me the difference in arrival time, or even overlaying the phase traces. Turn on the mains and set the measurement engine

BETWEEN THE LINES

delay set to the arrival, then bring on the front fill. This will cause a second peak in the Live IR indicating the front fill arrival. Hover your mouse cursor in the Live IR pane to see the delta delay time at the peak which represents the fill arrival, then mentally adjust that by 2 milliseconds, enter that value into the DSP as the front fill delay time and watch the Live IR peak snap into place 2 ms before the mains. This technique also has the advantage of doubling as a nice party trick on the rare occasions that you are tuning in the presence of another person who understands enough about using audio analyzers to be impressed by such things but hasn't read this book.

Now listen, walk, pay attention to the overall tonality of the system and adjust as needed. The subs will be louder towards the front of the room, which is another reason for setting the front fills a dB or two higher than target. This helps keep the tonality more constant from front to back, at the expense of some level variance. Temper that against the fact that listeners in the front row have self-selected and are expecting a more sub-heavy experience, and that is a better outcome than cranking the front fills and creating a sound exposure risk that is far in excess of the rest of the venue. The exact happy medium here may depend somewhat on the type of material being performed. Best to ask Acoustic Terry for a set list.

Theater L/R/Sub/Fill/Underbalc

Today's gig is in a larger venue: a 1200 seat theater with Left/Right line array mains, Left/Right subs on the deck in a three-per-side endfire configuration, front fills, and underbalcony delays. There is a balcony, but it's not been sold for this show, so we only must concern ourselves with the Orchestra level (floor seats).

Our mains are 8 boxes per side, driven in zones of 3, 3, and then 2 so we will refer those zones as A, B and C. We've designed our system to trim low enough to shoot under the balcony and make it back to mix position with 2 boxes of overshoot, then splays that give us center lines impacting evenly spaced from front to back across our coverage area, making sure we get down to the first row. Four front fills are driven off the same processor output, and all four of the underbalcony fills share an output as well.

We have three subwoofers per side and lots of space on the sides of the stage, so let's go for a three-element endfire array.

0. Verify
1. Main Left (level, shading and overall EQ)
2. Subs (endfire timing)
3. Subs (level + EQ)
4. Main L + Sub (combined response establishes target for rest of system)
5. Underbalcony fills (level and EQ)
6. Underbalcony fills (timing)
7. Front fills (level and EQ)
8. Front fills (timing)

Once we are verified, let's measure Main L at A, B and C (box 2, box

BETWEEN THE LINES / 151

5, and between box 7 and 8). It's a relatively short line so we can't do much about what's happening in the low frequency, but we can work on the high frequency shading until we are satisfied with the uniformity. 6 dB front to back would be a reasonable design goal in this situation so let's see if we can hit that with shading.

While we're here, let's use overall EQ to hit our target curve of choice - this array has a lot of tilt in it by default so let's flatten it back out a bit.

If we have three mics out, we can leave them at A,B,C for this series of operations and work quickly - and also leave them in place to examine the sub response at the next step. Remember the mic down front (C) is not in a very representative position (those bottom 2 boxes only cover a few rows) so let's be careful how much weight we give it in our decision-making process.

The endfire sub array will be easiest to time if we get a mic up close and personal so we can increase the direct:reverberant ratio of our measurement. We already have a mic down at C so let's just move that one until it's in front of our sub array and get to work.

Once we've finished the timing and have observed the expected summation in front of the array, we will adjust the sub level up or down until it lands against our desired target curve. It is probably best not to make this decision from the front rows - let's go further back in the room to get a more representative look at it. Then we will check alignment with the mains and observe the combined response, EQing as necessary. This gets stored in the analyzer to serve as our target response for the rest of the system.

The subsystem with the next largest custody area is the under-balcony delays, so we can fire one of them up (or all of them, since they're all on a DSP output together), measure one on axis (the one closest to FOH because it saves some steps) about halfway back through its coverage depth and then set its level and EQ until it matches our target response. We can apply some high pass or low shelf (or a combination) to these to minimize the low frequency energy bleeding off the back.

Now the delays create the same level halfway back in their coverage area that the mains make in theirs. We may "season to taste" depending on our application: do we want the same level all the way back, or do we want to let the level drop off towards the back? When setting levels, keep in mind that underbalcony fills may not be needed to restore the system to target level back there, but the "breath of fresh sound" in the form of higher direct to reverberant ratio is certainly helpful. Our analyzer will express its satisfaction in the form of improved coherence in their coverage area.

To make a timing decision once level and EQ have been set, turn on both systems. The concept of finding a seam is a little bit slippery here, since the area of overlap is large (in fact the majority of our underbalcony fill custody area is also covered by the mains). So we will set timing within the overlap area, but we need to decide where within that area specifically. If timing is set directly on axis to the delay fill such that the mains are arriving on time or early (I prefer to give the mains a 1 or 2 ms head start to keep localization towards the stage), the mains will remain early as we walk forward.

Conversely, in the situation where the underbalcony fills are covering seats that the mains can't get to, we have a more clearly defined "seam" towards the front edge of the underbalcony coverage, and it's clear where timing should take place.

We must remember that human perception plays an important

role in our timing decisions, so don't be afraid to massage it a few milliseconds one way or the other if it produces a more perceptually natural result.

Last on our decreasing-custody-size list is the front fills, and it's the same drill – with the fills off and the main system on, we can determine how deep the front fills need to cover, or in other words, how big the hole is. Halfway through that depth is a good on-axis measurement point, and that's where we'll set level and EQ to our target, with the option of being 1 or 2 dB higher due to stage sound. From there, we move backwards from the outer left front fill until we find its seam with the left mains, and time it to be 2 ms early at that location.

TV Studio (Distributed Mono)

This is a little more off the beaten path. It's a clamshell shaped TV studio with an audience area that's far wider than it is deep, and the seating area has a curved shape to hug the semi-circular stage platform. The system consists of three speakers hung along the edge of the stage – a left, a center, and a right, each covering a third of the seating section and only overlapping on the aisles. Since each listener will only hear one source, we need to drive this system in mono.

There's another ring of loudspeakers further back: a delay each for the far left and right corners of the seating area, plus a third hung on centerline to cover mix position, which is so far back that it's out of the coverage of the Main C loudspeaker. We also have a center subwoofer position comprised of two subwoofers sandwiching the Main C loudspeaker position.

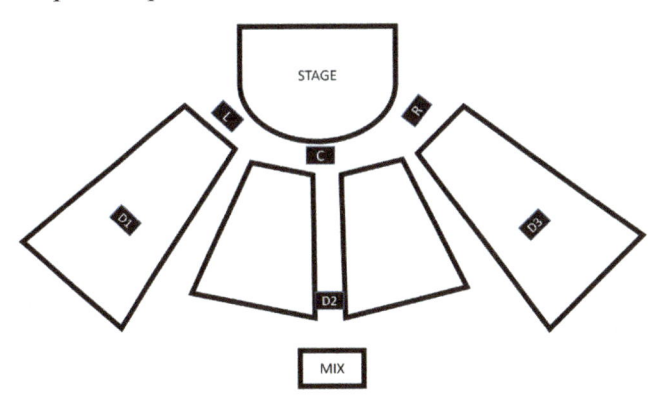

Figure 50 - The television studio, with L/C/R sources as mains and a trio of delays.

We can't rely on the muscle memory of our rock and roll L/R approach here, so this will be a good opportunity to use the Four Principles to guide us through the process.

The three main speakers – L, C and R – together make up the largest chunk of custody, so let's start with those, making sure each is doing its job within its custody area, and then we'll add the subwoofers to create our baseline target curve.

From there, we'll work outwards to fill in the delay zones D1 and D3 in the rear corners, and finally tune the mix position fill D2 to restore coverage at mix position to be as representative as possible of the response everywhere else in the room.

Our verification has confirmed that L, C and R loudspeakers are all properly matched. Left and Right are not aimed optimally - they overlap a bit with the C near the center of the room, and leave the outer edges of the seating section uncovered. Luckily they are on yokes so we can hop up on a ladder and aim them outwards a bit until we have "split the difference", achieving the same high frequency roll-off at the outer and inner edges of their custody areas (which conveniently land on aisles in this design).

Once we've sorted out the aiming, we start with Main C, grabbing a few measurements on axis front to back and evaluating vertical aim. We see that can't tip it up any more than it already is without the first few rows falling out of coverage, so let's leave it as it is. This system is used mostly for speech, so we'll tune to a relatively flat response.

Now we can move to the Main L and Main R and make sure they're keeping up their ends of the bargain. Same vertical aiming checks on both - adjusting the vertical tilt to cover as far back as we can while making sure that we're keeping the front rows in coverage - then matching level and EQ on axis for each source. Now we can find the seam between L and C and time it there, letting R follow along.

Next we turn on the subwoofers and adjust gain until we achieve the desired low frequency extension, not too drastic of a haystack for speech. Timing decision here is made about halfway out in the seating curvature, near the seam between L and C, which hopefully gives us a reasonably representative "average" location. and since it's two subs against three mains, the best approach is do the balancing and timing with everything turned on. We don't need to be too concerned about the multiple high frequency arrivals mucking up our measurement since there's so little overlap between the mains, but we do want to be able to take into account the low and low-mid frequency contributions from all the sources when they're all playing at once. This is also a good opportunity to capture a combined trace and store it as our full-spectrum target trace.

Onward to the delays! Since we tweaked the aiming of the mains, let's make sure the delays D1 and D3 are now aimed to cover their new custody areas. Then we'll set level and EQ on axis, and then move forward until we find the front edge of the delay and setting the timing with the mains there, making the delays 2 milliseconds late.

When we walk and listen to program material, we find that the delays feel too loud, likely due to the extremely reflective glass walls immediately behind the coverage area, so we turn them down by 2 dB and find that we like that much better.

Now that we know what the rest of the system is doing, we can bring our measurement microphone back to mix position and set level and EQ on the mix position delay fill D2 until it's representative of what's happening elsewhere, with enough delay to help the mixer localize to the mains. We must be careful here, since the mix position is not in the direct high frequency coverage of any of the mains, so we could easily get carried away with the delay and create a weird listening experience.

This also makes it more difficult to set an accurate measurement delay time, since the measurement delay finder won't be able to grab onto the peak in the IR caused by the presence of high frequencies. Smaart®'s Advanced Delay Finder allows us to lock our measurement delay into the arrival of a lower frequency from the mains (say, 250 Hz) and set the fill's delay time based on that. We will make our final adjustments based on listening to familiar program material.

Before we leave, we give the house mixer time to walk the room, get acclimated to the new coverage and learn how that relates to what's being heard at mix position.

Arena (Music Concert)

The arena system is basically a scaled-up version of the club and theater rigs. We still have Main L and R, and this time our subwoofers are flown L and R as well, behind the mains. The subwoofer array is 9 elements in a vertical line that acts to create some directionality in the vertical plane. Each main hang is 14 boxes over three downfill elements.

We have side hangs as well, to cover the wraparound on the sides beyond what the mains can cover adequately, and those are 12 elements each. Front fills outline the stage and a thrust, and front fill subs on the ground to restore the low frequency coverage in the front rows, where the listeners are standing underneath the main sub array.

0. Verify
1. Mains On-axis A,B,C,D for shading, level, EQ
2. Subs @ same locations for timing, level, EQ
3. Mains + Subs = Target
4. Side hangs on axis for shading, level, EQ to target
5. Side hangs + Mains + Subs for combined LF interaction to target
6. Front fills On axis for level, EQ
7. Front fill timing at seams

Both the mains and the side hangs have been designed by following the same rules of thumb: two boxes of overshoot, evenly spaced seating impact of center lines, and making sure we're getting down to the front.

After we verify, we'll pick one side of the Mains, make logical link groups in the DSP based on our seating area geometry (A = 200 level, B = 100 level, C = floor, D = downfill) and measure from there, where we can address high frequency shading and EQ to desired tonality.

Next, turn on the flown subs on the same side, adjust the overall gain until we get the proper extension to fill out our target trace down to the lowest frequencies, and then we can address the sub timing. Getting data of them on-axis means heading all the way up to 300 level, and it's a real challenge to get clean low frequency data from that far away, especially in a reverberant environment like an arena.

Ideally, we can set our alignment delays when the arrays are still floating off the ground, before they are flown to trim height, so it's easy once the system is in the air to simply confirm the alignment. We will observe the resulting summation, using EQ to remove any summation buildup that takes us above target. If we can't get the alignment ahead of time, the answer is, of course, do the best we can. Even from the far field, it should be simple enough to confirm that turning both sources on creates addition through the crossover region.

Now we head over to the sides, where we're working in two DSP zones - A in the 200's and B in the 100's. We address high frequency shading between the two zones, then feather the high frequency filters for a few boxes across the transition zone since we have the DSP granularity to do so.

We then EQ the overall array response to match our target above 200 Hz or so. There is going to be a lot of energy from the mains and subs spilling over into this area in the low frequency range, and we don't need our side hangs to contribute any more to it. This is the aforementioned 360° tuning strategy first introduced to me by

FOH engineer Jim Yakabuski – paying attention to what's coming off the sides and rear of our systems, not just the front.

We can use a low shelf to suck out some low frequency from the side hangs, and then bring the whole system on together to watch the response fill back in through the bottom octaves. Bump that low-shelf up or down a bit to land back at target on the sides. The additional advantage here is that we've reduced a lot of the low frequency energy from the side hangs that would be spilling onto the stage and back out front into the custody area of the mains, making the entire system better behaved in the low mid range.

We then need to address the timing between the main hangs and the side hangs. In reality they are hung in reasonably close proximity so we are not likely to need a very large amount of delay at the seam where they meet. This main-sidehang seam is a traditional weak spot for coverage in arena designs. Hopefully we have delta plate rigging on our side hangs so we can adjust the horizontal aiming once it's flown, and make sure that we're not leaving a gap.

Now we head down front to do our front fills – each one gets set to target level and tonality, and the level is a bit different for each fill due to the thrust geometry placing them different distances from the barricade.

The fill on the outer edge of the stage has most overlap with the mains (particularly the downfills) so let's set timing at that seam, leaving the fill to be 2 milliseconds early for imaging. That fill overlaps with its onstage neighbor, so we go to that seam next and set timing there. Finally, the fill on the tip of the thrust doesn't overlap substantially with any of the other side fills, but if we walk sideways in its custody area, we do start to get into the coverage of the bottom boxes of our mains, so let's find an appropriate overlap point and time there.

Day to day, the fill timing will remain consistent because they are attached to the stage geometry. The only thing we need to check is the timing of the entire fill system against the flown mains, because the system's trim height might be different. We therefore only need to spot-check the delay time where the outer fill meets the mains, and if we have the ability for our DSP to do multi-select and relative change, or we've got a tiered zone set up so we can operate on all the fills together, we just change the delay time once and the entire fills system follows along.

Finally, our ground subs. They restore the low frequency response for the first rows of listeners along the thrust, plus a little extra to give the expected tactile sensation of being in the front row at a rock concert. Walking backwards until we start to sense the energy from the flown subs is a good way to find the "seam" and set timing there. As with the delays in our theater example, the concept of a "seam" here is tough because there is a large region of overlap. However, we know exactly where the intended custody area is - the folks down front, along the thrust, out of the coverage of the flown sub array. So let's set the level until it's appropriate for those listeners, then back up until we feel like we're in the coverage of both the flown and ground subs, and make a timing decision there.

Arena (Installed System)

Now it's time for a game of sportsball. This 13,000 seat arena has a newly installed sound system to be used for music playback and announcements during hockey games, and it's our job to tune it. The drawings of the space we were sent aren't current, so when we arrive, we ask to be taken on a walk around the venue and have all the loudspeaker positions pointed out to us. As we walk, we build a mental map and approach by considering the Four Principles.

The system is symmetrical, with three loudspeaker clusters down each of the long sides (3,4,5). Each of these clusters consists of an upper box to cover the upper seating section, a lower box to cover the lower seating section, and a rear-facing box to cover the ice itself. The outer two clusters (3 and 5) also include subwoofers. There's another identical cluster at the North side (1), and the upper North bowl is covered by a line of five delay loudspeakers hung from the upper steel (D1 – D5). These are important because the energy from the main cluster is physically blocked by the building structure so the delays are the only coverage the upper bowl listeners will get.

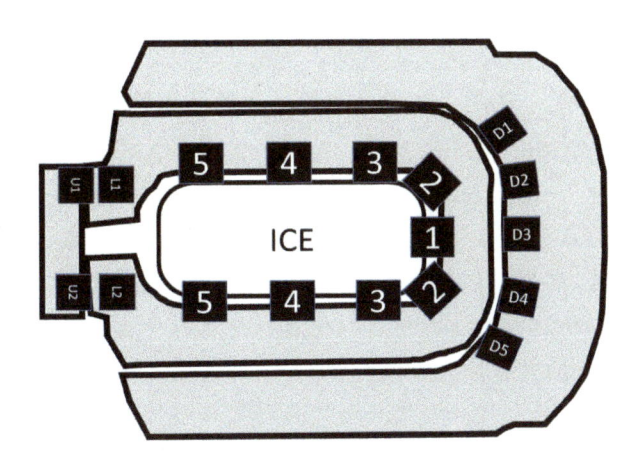

Figure 51 - Seating geometry layout and loudspeaker locations

There is an additional fill on the 45° corners (2), to fill in the gaps. The south end has **VIP** boxes in two sections – lower (L1/L2) and upper (U1/U2) – which don't have any high frequency overlap with each other, or with the clusters spanning the sides of the arena. Got your order of operations yet?

Work starts by having the freewire **DSP** file restructured to allow us the ability to adjust Gain, Polarity, EQ and Delay on each loudspeaker cluster, and additionally on each individual loudspeaker within each cluster. We will need both.

The lion's share of the seats are covered by the six clusters that span the long sides of the arena (3-5), so our principles tell us we can begin with one of those, and copy the settings to the rest after we're verified they're all matching. We start with the middle cluster on the West side (4), which is the cluster nearest to our control booth.

The top element is throwing the farthest and so will be the **SPL** and high frequency bottleneck, so let's start there. Since it's our farthest throw, we will want to bring back a little high frequency lost to the air absorption, but not so much that that these top elements will go into limiting far before the rest of the rig.

How far to go? Since we're not loudspeaker designers, this can be difficult to determine. We need to consider the operating **SPL** of the system, and how much headroom the amplifiers have above that before they hit limiters. If the system is likely to be driven to its full capabilities, we might choose to err on the side of less shading, whereas systems that will not be driven quite as close to their operational limits can be shaded more aggressively. In this case, I consulted with the loudspeaker manufacturer directly to gain some insight into this decision. Most manufacturers have application support available for situations such as this – don't hesitate to contact them.

We move down to the second box, covering the lower half of the audience seating (Onax B). Match level and response. Now we must glue A+B together, so we find the seam between the two, which the designer has thoughtfully landed on an aisle, and set timing there. Since the rear-facing ice speakers aren't always on, we will consider them a fill system and tune our system to hit its target response without them on for the time being. Copy those settings to the other five clusters and we've covered about 9000 seats.

Let's see what those subs are doing – they are a cardioid model so they won't be spilling across the arena to the other side, but the coverage is wide enough that all three clusters on each side don't need to have subs – just the outer two. So let's add the subs in at one of the clusters, setting level to hit target, then timing, then observing the summation, and copy that down to the others.

Next we go out to the North-facing cluster (1). Its geometry is the same as the clusters on the sides, so we're just making a quick spot check to make sure it's performing as intended. Then we have the corner fills (2) that connect the North cluster to the long sides. The installers didn't get the aiming quite right, so it's leaving about 80 seats a little light in the high frequency range. A mechanical problem (bad aim) requires a mechanical solution (fixing the aim), so we won't chase it with EQ but we will make a note of it for the client. In the meantime, it's a fill so we'll set its level and response to fill the hole, then time it at the seam(s). Oh, now we have a decision to make.

Do we time it to the main clusters along the sides (timing 2 to 3) or do we time to the end cluster (timing 2 to 1)? The physical geometry alleviates the choice here because both timings would result in the same delay time setting within a millisecond or two. If the difference was irreconcilable, we might time the fill into the sides (2 to 3), then time the North cluster into the fill (1 to 2), so everything is in time with its adjacent neighbor, and this is a rare situation where it makes sense for a Main source to have its timing dictated by the fill.

The next largest subsystem is the North upper bowl (D1 - D5). Tuning all five to the same level and tonality is quick work. As we walk and listen, we end up taking the level down by 2 dB because the listeners up here are so isolated from what's going on down on the floor, it feels too loud otherwise.

This would be especially true once the sportsball game is in progress and the folks up here in the bowl are not immersed in audience noise to the degree that the lower-level audience members are. These are the "quiet seats." We also high pass these delays by a significant amount, because the sub and low mid energy is making its way up here just fine, and in a venue this large and reverberant, the fewer sources we have dumping low frequency energy back into the space, the better.

We have yet another timing decision to make, and this is an interesting one. We have five delays that span the bowl horizontally, and two options:

We could choose to time each delay back down to the North cluster and the corner fills on the lower level (D1 to 1, D2 to 1, D3 to 1, etc). This would serve the folks sitting in the bottom few rows of the North bowl, who are far enough below the structural ceiling elements that they would be in the coverage of one of the delays and the lower-level mains. The downside is that the people seated on the seams between any two adjacent delays would have imperfect timing.

Alternatively, we could choose to time the five delays to be a smooth handoff horizontally all the way across the bowl. This would serve the folks who are seated on the seams between any two adjacent delays, but would have imperfect timing for the folks farther down in the bowl, on a seam between the mains and a delay.

The system designer has made the choice easy for us. The seam between the upper delay ring and the lower main system falls mainly on the ~~plain~~ mainly on the aisle. There are a very small number of listeners who will hear both the main system and the delay as well at high frequency. For the most part there is a lot of isolation in the delay system's custody area (which is why it was needed in the first place!) and those folks can't even see the main lower-level cluster, let alone hear it.

On the other hand, a lot of people are sitting in the overlap between two adjacent delays, so we should make sure those have a smooth transition. We time the middle delay (D3) at the bottom edge of its coverage, where it overlaps slightly with the main North cluster. From there, we time delay D2 into delay D3 at the seam, and then we work our way out and set timing between D1 and D2. Delays D4 and D5 followed along because they are symmetrical.

At this point the only thing we have left is "small potatoes" - the VIP systems (U1/U2/L1/L2). They're both isolated from each other - and from the rest of the system - so once we hit target level and tonality, we could very well consider the job done. But we are perfectionists so let's use the Advanced Delay Finder in Smaart® to time the 250 Hz energy in the VIP fills to the 250 Hz energy coming off the mains.

Let's not forget about those Ice fills! We (carefully) bring our mic out onto the ice, bring the fill up to target level minus a few dB (the crowd is screaming, hopefully the players are not) and then low-shelf out some of the low-mid energy so as not to contaminate the stands with it when the ice fills are on - we're getting more than enough off the backs of the main clusters as it is.

"Shed" with Lawn System

Ah yes, the indoor/outdoor amphitheater, affectionally known as the "shed." This type of venue is loved by mix engineers throughout the industry for its excellent acoustic qualities. After all, what could possibly be better than mixing a show where half the audience is inside of a highly reverberant metal structure, and the other half is outdoors?

We're going to treat this like two systems. We'll get the inside system tuned like we would any other venue, and then once we're happy with that, we'll extend the coverage to the lawn using the lawn delay system.

Mains are 14 elements per side of a large format array plus three flown subwoofers alongside each hang. There are four front fills across the front of the stage, plus a pair of point source boxes on the corners of the deck that cover the outer edges of the floor seating, which happen to be blocked by the large video screens on either side of the main hangs.

There are an additional six point source fills mounted atop the proscenium to cover the balcony seats, processed in pairs (inner pair, middle pair and outer pair). Let's sort out the indoor system first. After verification, we measure our Main L at three mic positions (A up in the balcony, B in the rear half of the floor and C down front) for variance and target curve. We then add in the subs while our mics are at the same locations.

Now that we have the target response established, we can tune the deck fills, which cover the outer edges of the seating area blocked by the flown video screens. The fills restore those seats to target response above the low-mid, but we'll take the level down by a dB or two to prevent anyone seated down front from being decapitated by the horn.

Next are those fills upstairs atop the proscenium. The outer pair is the most critical, because they cover seats that are mostly occluded from the mains by architectural features (how thoughtful) so they must have sole custody above the low-mid range to restore the target response in those seats.

The mid pair and the inner pair are covering balcony seats already well covered by the mains, so they act as a slight high frequency refresh and get turned down a significant amount because they also bleed a lot of off-axis energy into the rest of the house (we listened to them from down at mix position and didn't like what we heard). We want just enough level for them to put a little direct high frequency frosting on top, and no more.

Finally down to the front fills, and we know the game now – same level and tonality above the low-mid range, and we'll set timing at the seam between the outer fill and the bottom of the mains, since that's where the most overlap is. Then we walk and adjust to taste until we're happy with it.

Now it's time to tackle the lawn system. It consists of six bays, each of which has three loudspeakers inside: a main, a downfill box, and a very narrow horn mounted higher up to throw high frequencies up to the top of the lawn. All six bays have the same configuration, so we can align the three sources once, then copy and paste the tunings and timings to the other 5 bays. At that point we'll simply time each bay back to the main system underneath the roof.

The middle box of each cluster is the majority shareholder both in terms of how much of the lawn it covers and in terms of frequency bandwidth, so let's start by measuring that on axis and tuning it to the same response as our target curve from the indoor system. Next up is the downfill - match the response and level, then bring them both on, find the seam and time it. It's a little harder to find an exact crossover here, as the mounting inside the bay and behind the baffle treats us to a wide variety of interesting reflections and resonances, but by listening for the comb filter interaction we can find a location.

Once they're timed there, we turn on the top element - the HF catapult - and feather it in until the HF extension sounds as consistent as possible as we walk backwards across the lawn. There's no hard cut-off for end of coverage, so we just let it go up as far as is reasonable and trust that the folks waiting in line for hot dogs and beers care more about their snacks than a high-fidelity listening experience. After all, it's not our design...we're just tuning it.

The final step is to time each bay back to the indoor system. We are symmetrical again so we only must make timing decisions for bays 1, 2 and 3, knowing that 4, 5 and 6 will follow along.

We are technically in 100% overlap because the indoor system doesn't magically stop once it gets out onto the lawn, but it definitely needs help. So we need to determine a logical place to make the handoff and timing decision even though there's not a clear seam. The solution I used was to turn both the indoor and lawn systems on, and walk forward until I felt like I was on the bottom edge of the coverage of the lawn system. Backwards from this point, folks would be well in the coverage of the lawn bays, and forward from this point, they were getting 100% indoor mains. Right on this edge is where I set the timing and let the indoor system lead by just a handful of milliseconds, walking the seam again to ensure a smooth and natural sounding handoff.

CHAPTER 18
Controlling the Variables

No discussion of system optimization would be complete without addressing *that which cannot be optimized.* In every situation where you are called upon to tune a sound system, there will be factors you can control and factors you can't. In other words, let's talk about the real world.

All of the methods discussed in this book assume that we have control over the way the system is designed and deployed – product selection, box counts, point locations, trim heights, splay angles, available processing, tuning time...

It's a natural reaction to look at a large-format PA system and think "well I could get good results too if I had unlimited budget." If there is a gig with unlimited budget, I haven't gotten the call for it. Hopefully the preceding examples have convinced you that my approach is the same regardless of the scale, scope, or complexity of the system, and so my process is always applicable. However, there are always practical limitations of some sort.

For example, with a touring act carrying its own PA, we would have control every day of how many loudspeakers to hang, trim heights and splay angles, and how it's driven, and we are able to design a system every day that best covers the space. But we're not able to control the total amount of PA that the production's budget can accommodate, or the sightline limitations, or the rigging point locations used by other departments (lighting and video). Maximum achievable trim height and number of boxes will be determined by the rigging restrictions at the venue each day, and we need to work within that.

If we're guests working on an installed system in a club or theater, we probably don't have the ability to change trim heights or splay angles, but we can probably tune, shade, and time it as we see fit. Festival situations can run the gamut - acts performing earlier in the day shouldn't expect to have much in the way of control over the system past being able to adjust the overall EQ to achieve the target tonality. Headlining acts can likely be more involved with both the design and tuning of the system.

What we have here is a spectrum, with full control at one end and a simple layer of overall EQ at the other. Every event will land somewhere on that spectrum. So what can we do to get as much control as we can over as many of the variables as we can so we can optimize the final result?

Getting involved as early as possible is a key component. The sooner we are involved in the planning process, the more likely it is that we can have input on gear selection, loudspeaker positions, quantities, aiming, and other design factors. No amount of EQ will fix a bad design. We can't expect to use our processing to compensate for a poorly-aimed loudspeaker. If we get involved with the design early, we can resolve some issues before they become issues.

Let's not neglect the role of the human in this work. We want to have a good, productive relationship with all parties involved. Hopefully everyone sets ego aside and is willing to work together for the good of show. Upon arrival at a new site, I will spend some time speaking with the other folks. If I'm working for the artist, I'll spend some time chatting with the vendor techs and crew. If I'm working for the vendor or venue, I'll ask the touring audio crew about how the tour has been going.

Remember people's names, be friendly and professional, and start your day off on the right foot. These are relationships that will last much longer than a single event and we want to cultivate them, as well as a positive professional reputation. If people like us, they will be more willing to want to work with us. It's a small industry – we will see these people again!

On a related topic, it is never a good idea to surprise anyone. The advancing process is very important. When I work for artists, I work with production management to ensure that our advancing materials includes a lot of information about our expectations for the sound system design, deployment, and processing, what we are bringing, what we expect them to supply, and any other relevant information so everyone is starting off from a place of understanding, and not being surprised with it.

Some of the things that I would typically request in a rider (advance approval of predictions, no more than 2 elements per processing zone, no 0° angles, specific processing preferences, etc) are straightforward if addressed at the design stage, but major complications if I wait until show day and then show up and ask for it. Starting communication early increases the odds that we will get what we are asking for.

As soon as you are able, I recommend investing in your own front-end processor and basic measurement kit. A laptop, interface and measurement mic can easily fit in a backpack. This means you always able to measure any system you walk up to and are never reliant on anyone else to be able to gather information and help yourself make more informed decisions.

A processor can be a more substantial investment, but having the ability to set your own overall EQ (Tier 1) and work on each system zone (Tier 2) means you can make a lot of headway towards uniformity without needing access to the vendor's processing or the installed system. If we're going to hand over Left/Right/Sub/Fill feeds anyway, let's have a processor inline so we can work on them and get the system to our liking before the signals leave our custody. The only thing left to do is implement shading and loudspeaker presets in the system's dedicated processing.

There is an added benefit to carrying your own processing: speed and efficiency. This is the same reason mix engineers request a specific console even though many mixing consoles have similar functionality - they are comfortable with a specific platform and tool which allows them to work faster and more efficiently and thus consistently produce a higher quality result. The same is true for processing - find a platform that you can operate quickly and comfortably, and that minimizes the amount of time you need to spend navigating the wide variety of platforms and interfaces that are currently in use for system processing and control. Making sure I have the processing available to do my job is very much worth the $30 it costs me to baggage-check a flight case containing my DSP of choice.

I often hear from folks who are still early in their system engineering journey, and operate in small markets with very limited technical tools, and typically smaller events where rider requirements are not in play. Vendors in these markets are less likely to be operating with modern equipment and can be dismissive or outright hostile towards notions of system measurement, prediction, or optimization.

At the end of the day, there's only so much you can do, but many of these folks have had success by bringing their own analysis and processing tools to bear to show the "powers that be" what improvements are possible, and then it's typically an easy sell.

In situations where we have very limited control, our day can become frustrating very quickly. After all, how are we supposed to do our job without access to the proper tools?

In this sense, our view of system optimization becomes rather pragmatic - do the best job with what we have. If all we have is access to processing, but the system needs mechanical changes, we simply must be realistic about what improvements we can make, make those, and then stop beating ourselves up about factors we can't control.

The figure below shows a medium-format array that is permanently hung in a theater with a large balcony. Unfortunately, the overall aiming of the array is improper. There are no loudspeakers covering the balcony. The top pane shows the response with all processing bypassed, and the bottom shows the response after the manufacturer's proprietary beam steering algorithm is engaged. It's done a reasonable job of high frequency shading throughout most of the venue, but is unable to make much improvement in the balcony (green trace) simply because loudspeakers aren't aimed there.

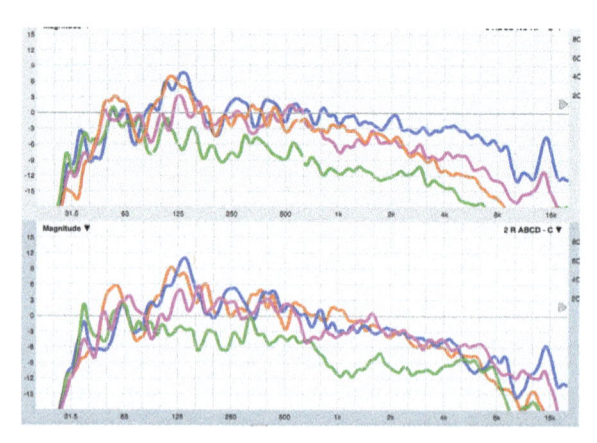

Figure 52 - A poorly designed sound system, without (TOP) and with (BOTTOM) the manufacturer's proprietary beamsteering processing engaged. Since there are no loudspeakers covering the balcony (GREEN TRACE), the processing isn't able to do much to help. The solution here is to take down the system and rehang it with proper aim. (Smaart® v8)

In this situation, if we are visiting the venue and don't have the ability to make physical changes to the loudspeaker deployment, the balcony will simply be a case of "it is what it is."

Get the rest of the coverage as uniform as possible, and tell our FOH engineer that we've done everything we can without landing the system and changing the rigging. We can continue to be helpful by walking the coverage and informing the FOH engineer of the variance so they can attempt to take it into their account in their mix if they so choose.

There's no shame in saying "I've done the best I can with what I have." In fact, that's the *entire job*. Knowing what we should spend time trying to fix, and what we need to come to peace with, is a critical skill to have here - both technically and politically speaking. Do the best with what you have - focus on the things we can improve, and stop worrying about the things we can't.

Defending the Show

As sound system engineers (or at least as individuals acting in that capacity), we are the custodians of the mix once it leaves the console and the sonic experience of the show all the way to the audience. We can take ownership of the fact that we are in a unique position, with a unique perspective on how the sound of the show is being delivered to the listeners. It's common to encounter deployments where there aren't enough front fills, or out fills, or delays, or a lawn system, or some other oversight that leaves listeners uncovered.

In these situations, we must remember that if we don't say something, it will probably go unaddressed. There are a lot of forces in a modern production environment that work in the direction of fewer loudspeakers, hiding loudspeakers in inadequate locations, taking away front fills, saving money by not having side hangs, and so forth. We are the check and balance that opposes those forces and helps to make sure that the audience gets to hear the show they came to see (and in many cases, paid to see). Odds are that if we don't advocate for the proper coverage, it won't be improved.

That's not to say that we'll always get what we ask for. Production management may say "no." It may not be in the budget. But the responsibility still falls with us as system designers and engineers to make sure that we are doing what we can do to advocate for the artist's vision and the sonic experience of each audience member. This is particularly true if the artist is signing our paycheck. That paycheck signifies the trust that the artist has placed in us to act on their behalf, and to represent their interests in terms of how their show sounds and is experienced by the audience. This responsibility should not be taken lightly, and we should remember that we are directly representing the artist as we conduct our business, and act accordingly.

After all, any audience member in any seat can film the show from their phone, post it to social media and tag the artist before the show is even over. When your artist asks you why a YouTube video from their performance has thousands of views and an inaudible lead vocal, what are you going to say? Defend the show, and advocate for the tools you need to do your job and cover the listeners, so you never have to have that conversation. In other words – if we don't say something, who will?

Play the hand you're dealt.

CHAPTER 19
Common Pitfalls

When I mentioned to my system engineer friends that I was working on this book, this chapter was the most suggested idea – a collection of common mistakes, misunderstandings and pitfalls from the system design and tuning process that we've seen repeatedly.

By far the most widespread issue I've observed with modern sound system engineering is that far too many people make decisions or perform operations on a system without thinking about why they're doing it. This lack of intent is something I'm hoping to address head-on with this book, because everything else stems from that.

Underspecified Systems

Another very common issue is under-specified sound systems. I would go so far as to say that most sound systems are under-specified, which has the unfortunate effect of making a properly specified system seem like overkill by comparison. The painful truth is that a lot of the people who specify sound systems are account managers, production managers, sales reps, or front of house engineers, not systems engineers or qualified system designers.

Underspecified Mains

At small scale events, underspecified mains often take the form of "speakers on sticks," but a loudspeaker that's underpowered for the application frequently inspires attempts to "make it louder" by adding a second loudspeaker, fully overlapped with the first, which of course only stands to give a broadband SPL increase of +3 dB (twice the power) at the expense of a savage comb filter, which destroys intelligibility and uniformity.

Figure 53 - LEFT: A single loudspeaker at 4 kHz. RIGHT: an attempt to "double up" creates a coverage disaster. Just say no! (MAPP XT)

If we need higher SPL in the listening area, we either need a more powerful loudspeaker, or we need to use loudspeakers that are designed to be arrayed to couple and create a higher SPL into the coverage area (i.e. a line array).

With flown line arrays, under-specification of mains means that we either can't adequately cover the entire vertical coverage angle required, or we have barely enough boxes to do so by using very wide splays. This robs us of the ability to use progressive splays to reduce the level variance. If we have a 45° vertical coverage angle, and only three 15° boxes to cover it, must open all our splays to the maximum 15° to cover all the seats.

If we have more than three boxes, we can adjust our splays in a strategic fashion to send more output towards the furthest listeners. In other words, it seems that all too often, box count is determined simply by estimated **SPL** needs, ignoring the factors of vertical coverage angle and range ratio. Adding a few more elements to a flown line array is one of the most effective things we can do to equip ourselves to achieve a higher level of uniformity, both because of the additional splay angle flexibility, and because of the increased line length.

Underspecified Side Hangs

Underspecified side hangs are another common issue. For some reason there seems to be a pervasive belief that side hangs should be significantly smaller and less powerful than the main hangs. In some cases, this is okay, particularly when the side hangs only need to cover a small area to a small depth, but in a more typical deployment such as an arena or amphitheater this falls flat.

In these types of environments, the side hangs typically need to cover a vertical coverage angle that is at least as large as the mains, and will pick up coverage in the horizontal plane where the mains left off, which means they need to throw almost as far.

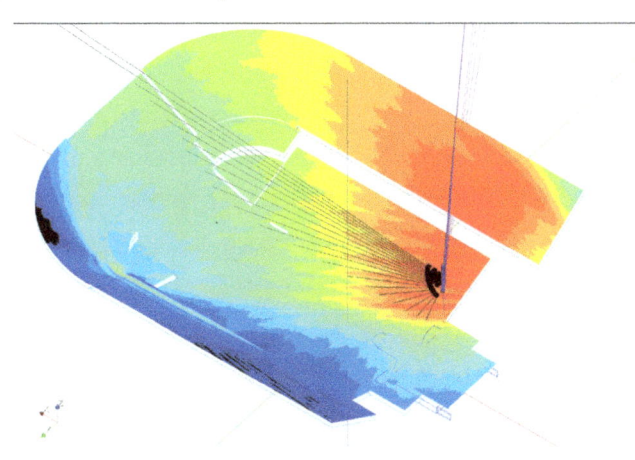

Figure 54 - A side hang in an arena, predicted at 4 kHz (MAPP 3D)

The figure shows a side hang in an arena predicted at 4 kHz. The coverage shape is highly asymmetrical in the horizontal plane, with the left side of the box throwing much further than the right. If we are only considering the on-axis response, it might be difficult to justify a long line of powerful boxes, but when we see how it needs to get up into the far corner without severing the heads of the folks listening to the bottom elements, we can see that a longer line with more angles to articulate and more elements to shade will give us more flexibility and higher success rates when dealing with such geometry. As a rule of thumb, I want at least 75% as many boxes in my side hangs as I do in my main hang (20 mains / 16 on the side, or 16 mains / 12 on the side).

Underspecified Subwoofers

A higher number of subwoofers increases our ability to achieve better directivity and uniformity by allowing us to create more complex arrays made of cardioid elements and so forth. Besides the directivity, there is a more subtle benefit to more subwoofers.

Since human hearing is less sensitive to lower frequencies, our ears tend to emphasize the effect of distortion harmonics produced by subwoofers, since they are higher in frequency and thus exaggerated by our ears compared to the fundamental tone that produced them. For this reason, a few more subwoofers per side allows us to achieve the same **SPL** but without each individual subwoofer working as hard, which keeps the subwoofer array more linear with less distortion. This leads to a "cleaner" sound. If you typically can get the required **SPL** with two subs per side, for example, try using three per side and see whether you notice an improvement in sound quality.

Underspecified Front Fills

We continue our survey of underspecified system elements by

considering front fills. This one is short and sweet: for the folks down front, out of the coverage of the mains, the front fill must work alone and bear the entire brunt of SPL duties for those listeners.

For many types of event, it makes sense for the front fills to be a dB or two higher in level than what the mains are achieving back in the house, so it follows that we want a powerful fill speaker that is able to achieve show level on its own without distortion or limiting. As a rule of thumb, if you are worried about protecting your front fills with a limiter, they are probably not powerful enough for a typical live music event.

In a proper low-variance sound system design, the front fills do a lot more work than you might think, and all too often they hit limit before the main PA, creating quite an unpleasant listening experience for the folks down front (who probably paid a lot of money for those seats). Front fills are the sole custody holder for many of the biggest fans, so let's make sure we have the right tools for the job.

Front Fill Depth

Another closely related pitfall here is assuming the front fills will penetrate the listening area further than they actually will. If the front fills are not above head height, they are only good for a row or two, period. Beyond that, they are blocked by human heads. Only on larger, higher stages can we reasonably expect front fills to penetrate further.

Since the listeners are so close to the source, level drops very quickly as we back up thanks to the inverse square law. So giving front fills a deep custody area typically leads to them being too loud up close and not loud enough further back. It's best to design the main system to cover as far down to the front as possible, and only rely upon the front fills to cover the first row or two.

Lack of Overshoot

This is a common offender in installed systems where the designers are overly fearful of the back wall of the venue. If the top box is aimed directly at the last row of listeners, they are outside the high frequency lobe and will experience inadequate coverage at high frequencies. Overshooting by a box or two helps bring those listeners back into a more representative part of the coverage.

Remember the "smiles and frowns" issue - depending on the venue geometry, if the centerline of the top cabinet is aimed perfectly into the last row of seating, the coverage further out towards the edges of the box might be missing the party completely. As a rule of thumb, I start with two boxes of overshoot, but remember that prediction will show us how we are doing, so no need to guess.

Random Microphone Positions

There are many ways to skin this cat, but we must remember why we are measuring (and what we are hoping to learn) and choose a microphone position that will yield the relevant information. The zoning strategy described in Chapter 16 allows us to create a direct correlation between the HF shading needed, and the boxes we need to operate on to address it. When mic positions are arbitrarily chosen, even if issues are observed, there is no clear indicator of where in our DSP we need to go to address it. Observing a parity between listening area architecture, array DSP zoning and microphone positions makes these types of decisions very straightforward.

Level and EQ are set from within a loudspeaker's custody area, so that's where the mic should go. Timing decisions are made at the seams between sources, so that's where the mic should go.

0° Angles

As discussed previously, 0° splays are often used in an attempt to get the array to "throw" further. A 0° angle creates a high frequency coverage shape that narrows over distance (and over frequency) and is a tradeoff not worth taking in my experience. Either way, coverage is going to fall off towards the back, but with proper splays, it can least maintain some semblance of tonal consistency as it falls off and transitions to the delay system. If you need delays to obtain sufficient coverage with your mains splayed at 1°, you will almost certainly still need the delays when splayed at 0° and you've just made things a lot harder to optimize.

Laser beam Centerlines

Don't fall prey to the belief that line arrays only emit sound along the centerlines - we discussed the pitfalls associated with trying to skip balconies and other architectural features with large inter-element splay angles. Besides the geometric realities (it doesn't usually work very well), a single large splay causes high frequency disruption that decreases uniformity within our coverage area, and it's usually not a tradeoff worth taking.

By the same logic, don't give the centerline locations more weight than they deserve. There is a prevailing belief that splays should be adjusted so FOH mix position doesn't land on the transition between boxes. Bad news - if you are standing in front of a line array, you are listening to transitions between boxes. While I will work to prevent mix position from being covered by a seam between different *products* (main-downfill transition, for example), we must remember that even at maximum splay there is some amount of overlap between cabinets, so be mindful of how much energy and time you dedicate to trying to avoid it.

Relying on Autosplay

Autosplay algorithms can be a helpful starting point for setting the array splays to the approximate required vertical coverage angle rather than starting your design from a straight array of all zeros. However, splay angles should always be adjusted until you are happy with them. Pay particular attention to making sure you have the proper overshoot, that your center lines are appropriately spaced and that your curvature makes it all the way down to the front of the listening area.

Skipping Verification

Murphy's Law is in full effect here. I see far too many systems where left and right don't match, the trim heights don't match, the angles don't match, the front fills don't match, some of the subs are of inverted polarity, and so forth. Trying to tune a system that hasn't been checked for these issues is likely to result in compounding those issues with additional issues. Verify, verify, verify.

Filter Myopia

Well-designed sound systems can be effectively tuned with a handful of wide, gentle filters addressing low-frequency tilt and high-frequency shading. If I encounter a system processor sporting a large number of deep, narrow filters ("picking carrots"), this is an indicator that something else is seriously wrong. Often, someone has attempted to EQ away a comb filter without realizing that's what they were doing. Severe EQ is an indicator that the mix engineer is chasing something in the time domain, which of course won't go away with EQ. Often what they are hearing is a disruptive reflection at mix position, or the late arrival of another system (front fills etc).

Our goal with equalization is to make effective improvements in tonality that hold across a source's coverage area. Smoothing and averaging are helpful tools to help us evaluate the general response trends and suppress the seat-to-seat micro-ripple.

The deeper and narrower the response deviation, the more likely it is to be a localized occurrence (that disappears or moves if we measure from a few seats away). These comb filters that cycle very quickly over the space are a result of larger time offsets between the contributing signals, which usually means we're seeing the results of interaction of physically displaced sources.

Trying to equalize such a spatially volatile interaction will be met with very limited success at a single point in space at the cost of degradation of the response everywhere else – not a bargain worth taking since our listening area holds more than one person. Focus on the general trends that are consistent over the coverage area, as those are prime candidates for effective equalization.

Tuning with Both Sides On

If you measure a PA system with both sides of the Mains turned on, you are measuring the combined response (a comb filter that is unique to that microphone position). Other measurement locations will yield different comb filters due to the varying time offset. Don't try to place 20 filters to EQ away a comb filter that only occurs at that particular point in space.

There are some cases in which we might want to observe the combined behavior of both sides, for example, if we need LEFT and RIGHT to sum with a mono subwoofer array. However, we must remain conscious of the fact that the combined interaction will change over the space and thus extreme caution should be observed before attempting to correct it with equalization.

Prioritizing Gear over People

As much as I care about the proper design and tuning of a system, it's important that I never let that become a higher priority than my relationship with the other professionals around me. We want to have a positive, respectful working relationship with everyone involved with the production. Is it worth asking the crew to drop and rehang an array because two of the splays are off by 1° from what I requested? Maybe they'd do it for me, but it wouldn't earn me any friends and I probably won't be able to get much else from them for the rest of the day.

Understand that professionalism and respect are the most valuable currency, and spend that currency wisely. In other words – choose your battles. We need to use our evaluations of priority to figure out what it makes sense for us to ask for, and what we should just come to peace with.

Every event is an opportunity to develop new friends and professional relationships within the industry, and those relationships will last far longer than the 90-minute event we've been hired to make happen. I strive to conduct myself in a way that will lead people to speak positively of me, my team and my work when the event is over. I want them to be happy to see me when our paths cross again. "That guy made us do half an hour of work for a 1° splay change" is not the reputation I want following me around. I'd rather have "he was cool to work with, we had a fun time together and his show sounded great."

Be polite, be reasonable and respectful, remember people's names, and be ready to make compromises where it makes sense to.

Conclusion

We've reached the end – for now. It took several years, hundreds of tuning sessions, and a bunch of mistakes for some blog posts to turn into the book you're holding in your hands. Although it is surely imperfect (visit my website at precisionaudioservices.com/book for a list of errata as they are inevitably discovered, as well as supplementary material), my sincere hope is that it has provided you with some help, insight, or at least a handful of thought-provoking ideas. Regarding this book, to quote console designer Douglas Self, "any suggestions for improvement that do not involve its combustion will be gratefully received." If you have made it this far, I hope you find success in your work and that you help to push for higher standards of excellence, raising the tide of our industry, and all our boats along with it.

Recommended Reading

McCarthy, B. ***Sound Systems: Design and Optimization,*** 3rd edition. Focal Press.

Rational Acoustics. ***Smaart® User Guide.*** RationalAcoustics.com

Further Resources

Boyce, Teddy. ***Introduction to Live Sound Reinforcement,*** 2nd edition. FriesenPress.

Davis, D., Patronis, E. & Brown, P. ***Sound System Engineering*** 4th edition. Taylor and Francis.

Davis, G & Jones, R. ***Sound Reinforcement Handbook,*** 2nd edition. Hal Leonard.

Everest, F.A. ***Master Handbook of Acoustics, 6th edition.*** McGrawHill

Huntington, J. ***Introduction to Show Networking.*** Zircon Designs

Lawrence, M. ***ProSoundWeb*** Author Archive. prosoundweb.com/author/mlawrence/

Rational Acoustics. ***Smaart® Operator Fundamentals.*** RationalAcoustics.com/training

ProSoundWeb. ***Signal to Noise Podcast.*** Hosted by Michael Lawrence, Kyle Chirnside, Chris Leonard and Sam Boone. SignalToNoisePodcast.com

SynAudCon. ***Course 130 – Equalization: A systematic approach to sound system tuning.*** ProSoundTraining.com

Winer, E. ***The Audio Expert,*** 2nd edition. Focal Press.